Computer Techniques
in Vibration

Mechanical Engineering Series

Frank Kreith and Roop Mahajan - Series Editors

Published Titles

Computer Techniques in Vibration

Edited by

Clarence W. de Silva

The University of British Columbia
Vancouver, Canada

CRC Press
Taylor & Francis Group
Boca Raton London New York

CRC Press is an imprint of the
Taylor & Francis Group, an **informa** business

CRC Press
Taylor & Francis Group
6000 Broken Sound Parkway NW, Suite 300
Boca Raton, FL 33487-2742

First issued in paperback 2019

© 2007 by Taylor & Francis Group, LLC
CRC Press is an imprint of Taylor & Francis Group, an Informa business

No claim to original U.S. Government works

ISBN-13: 978-1-4200-5317-3 (hbk)
ISBN-13: 978-0-367-38934-5 (pbk)

Library of Congress Cataloging-in-Publication Data

Computer techniques in vibration / editor Clarence W. de Silva.
    p. cm.
  Includes bibliographical references and index.
  ISBN-13: 978-1-4200-5317-3 (alk. paper)
  ISBN-10: 1-4200-5317-5 (alk. paper)
  1. Vibration--Data processing. 2. Vibration--Mathematical models. I. De Silva, Clarence W. II. Title.

TA355.C665 2007
620.3028'5--dc22   2006100170

Visit the Taylor & Francis Web site at
http://www.taylorandfrancis.com

and the CRC Press Web site at
http://www.crcpress.com

Preface

In individual chapters authored by distinguished leaders and experienced professionals in their respective topics, this book provides for engineers, technicians, designers, researchers, educators, and students, a convenient, thorough, up-to-date and authoritative reference source for computer techniques, tools, and signal analysis including finite element methods and wavelet analysis and the use of MATLAB® toolboxes, in the field of mechanical vibration. Important information and results are summarized as windows, tables, graphs, and lists throughout the chapters, for easy reference and information tracking. References are given at the end of each chapter, for further information and study. Cross-referencing is used throughout to indicate other places in the book where further information on a particular topic is provided. Examples are given throughout the book to illustrate the use and application of the included information. The material is presented in a format that is convenient for easy reference and recollection.

Mechanical vibration is a manifestation of the oscillatory behavior in mechanical systems, as a result of either the repetitive interchange of kinetic and potential energies among components in the system, or a forcing excitation that is oscillatory. Such oscillatory responses are not limited to purely mechanical systems, and are found in electrical and fluid systems as well. In purely thermal systems, however, free natural oscillations are not possible, and an oscillatory excitation is needed to obtain an oscillatory response. Low levels of vibration mean reduced noise and an improved work environment. Vibration modification and control can be crucial in maintaining high performance and production efficiency, and prolonging the useful life in industrial machinery. Consequently, a considerable effort is devoted today to studying and controlling the vibration and shock generated by machinery components, machine tools, transit vehicles, impact processes, civil engineering structures, fluid flow systems, and aircraft. Before designing or controlling a system for good vibratory or acoustic performance, it is important to understand, analyze, and represent the dynamic characteristics of the system. This may be accomplished through computer analysis of analytical models and analysis of test data, which is the focus of this book.

Clarence W. de Silva
Editor-in-Chief
Vancouver, Canada

Acknowledgments

I wish to express my gratitude to the authors of the chapters for their valuable and highly professional contributions. I am very grateful to Michael Slaughter, Acquisitions Editor-Engineering, CRC Press, for his enthusiasm and support throughout the project. Editorial and production staff at CRC Press have done an excellent job in getting this volume out in print. Finally, I wish to lovingly acknowledge the patience and understanding of my family.

Editor-in-Chief

Dr. Clarence W. de Silva, P.Eng., Fellow ASME, Fellow IEEE, Fellow Canadian Academy of Engineering, is Professor of Mechanical Engineering at the University of British Columbia, Vancouver, Canada, and has occupied the NSERC-BC Packers Research Chair in Industrial Automation since 1988. He has earned Ph.D. degrees from the Massachusetts Institute of Technology and the University of Cambridge, England. De Silva has also occupied the Mobil Endowed Chair Professorship in the Department of Electrical and Computer Engineering at the National University of Singapore. He has served as a consultant to several companies including IBM and Westinghouse in the USA, and has led the development of eight industrial machines and devices. He is recipient of the Henry M. Paynter Outstanding Investigator Award from the Dynamic Systems and Control Division of the American Society of Mechanical Engineers (ASME), Killam Research Prize, Lifetime Achievement Award from the World Automation Congress, Outstanding Engineering Educator Award of IEEE Canada, Yasundo Takahashi Education Award of the Dynamic Systems and Control Division of ASME, IEEE Third Millennium Medal, Meritorious Achievement Award of the Association of Professional Engineers of BC, and the Outstanding Contribution Award of the Systems, Man, and Cybernetics Society of the Institute of Electrical and Electronics Engineers (IEEE).

He has authored 16 technical books including *Sensors and Actuators: Control System Instrumentation* (Taylor & Francis, CRC Press, 2006); *Mechatronics—An Integrated Approach* (Taylor & Francis, CRC Press, Boca Raton, FL, 2007); *Soft Computing and Intelligent Systems Design—Theory, Tools, and Applications* (with F. Karry, Addison Wesley, New York, NY, 2004); *Vibration: Fundamentals and Practice* (Taylor & Francis, CRC Press, 2nd edition, 2006); *Intelligent Control: Fuzzy Logic Applications* (Taylor & Francis, CRC Press, 1995); *Control Sensors and Actuators* (Prentice Hall, 1989); 14 edited volumes, over 170 journal papers, 200 conference papers, and 12 book chapters. He has served on the editorial boards of 14 international journals, in particular as the Editor-in-Chief of the *International Journal of Control and Intelligent Systems*, Editor-in-Chief of the *International Journal of Knowledge-Based Intelligent Engineering Systems*, Senior Technical Editor of *Measurements and Control*, and Regional Editor, North America, of *Engineering Applications of Artificial Intelligence – the International Journal of Intelligent Real-Time Automation*. He is a Lilly Fellow at Carnegie Mellon University, NASA-ASEE Fellow, Senior Fulbright Fellow at Cambridge University, ASI Fellow, and a Killam Fellow. Research and development activities of Professor de Silva are primarily centered in the areas of process automation, robotics, mechatronics, intelligent control, and sensors and actuators as principal investigator, with cash funding of about $6 million.

Contributors

M. Dabestani
Furlong Research Foundation
London, United Kingdom

Marie D. Dahleh
Harvard University
Cambridge, Massachusetts

Clarence W. de Silva
The University of British Columbia
Vancouver, British Columbia, Canada

Ibrahim Esat
Brunel University
Middlesex, United Kingdom

Giuseppe Failla
Università degli Studi
Mediterranea di Reggio Calabria
Italy

Mohamed S. Gadala
The University of British Columbia
Vancouver, British Columbia, Canada

Cheng Huang
National Research Council of Canada
Vancouver, British Columbia, Canada

Zhong-Sheng Liu
National Research Council of Canada
Vancouver, British Columbia, Canada

Nikolaos P. Politis
Rice University
Houston, Texas

Datong Song
National Research Council of Canada
Vancouver, British Columbia, Canada

Pol D. Spanos
Rice University
Houston, Texas

Contents

1

Numerical Techniques

Marie D. Dahleh
Harvard University

Summary

This chapter gives an overview of numerical techniques for vibration analysis. The centered difference approximation for the first, second, and fourth derivative are given. These form the basis for the finite difference approximation of both spring–mass systems and the continuous problem. The fourth-order Runge–Kutta method is presented. Both of these approaches are used to solve the single-degree-of-freedom (single-DoF) system. In order to demonstrate these techniques for the multiple-degree-of-freedom (multi-DoF) system a two-degree-of-freedom (two-DoF) system is explored. Finite element and finite difference methods are presented as solution techniques for the continuous problem (also see Chapter 4). The bar and beam are used for examples. The Rayleigh method and Dunkerley's formula are presented. These are methods for computing the fundamental frequency.

1.1 Introduction

This chapter presents numerical techniques for three classes of vibration problems; the single spring mass, the multiple spring mass, and the continuous model. The simplest oscillatory problem results when the problem can be modeled by the single spring–mass system. This system is modeled by a second-order constant coefficient differential equation, which can be solved analytically when either there is no forcing, or the forcing is harmonic. When the forcing is not harmonic, the finite difference method can be used to solve this second-order equation or to solve the system of first-order equations, which is equivalent to the second-order equation. A larger computational problem arises when there are multiple springs and masses subjected to nonharmonic forcing. Again, the finite difference method can be used to compute the full solution. It is not always necessary to compute the full solution. Often, all that is needed is the fundamental frequency. This can be computed from the eigenvalues of a matrix problem. Both the numerical solution to the full problem and the matrix problem will be discussed. Finally, one may want to look at the vibration of a continuous element like a beam. Here, both the finite difference method and the finite element method will be discussed.

1.2   Single-Degree-of-Freedom System

The simple undamped spring–mass system oscillates with a frequency known as the natural frequency. The natural frequency is determined by the spring constant and the mass in the following way, $\omega = \sqrt{k/m}$. This relationship is derived by applying Newton's second law to the basic spring–mass system. The resulting equation is given by $m\ddot{x} = -kx$. The solution to this equation is harmonic with the frequency specified by the expression $\omega = \sqrt{k/m}$.

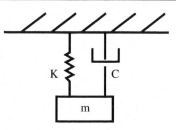

FIGURE 1.1   Single spring–mass system.

A more realistic vibration model of a simple oscillatory system includes a mass, a spring and a damper (see Figure 1.1). For simplicity, the mass is concentrated at the center of mass of the object, the spring is assumed to be of negligible mass, and, for the purposes of this discussion, the damping will be modeled by viscous damping. This is described by a force proportional to the velocity, denoted by $\dot{x}$. In the case of no external forcing, the system can be described by the following equation:

$$m\ddot{x} + c\dot{x} + kx = 0$$

where m, c, and k are constants and m is the mass, c is the damping coefficient, and k is the spring constant. This equation can be solved analytically by assuming a solution of the form

$$x = e^{st}$$

where s is a constant. Substitution into the differential equation yields the following quadratic equation:

$$(ms^2 + cs + k)e^{st} = 0$$

This equation is satisfied for all values of t when the following quadratic equation, known as the characteristic equation, is satisfied:

$$s^2 + \left(\frac{c}{m}\right)s + \frac{k}{m} = 0$$

The characteristic equation has two roots

$$s_1 = -\frac{c}{2m} + \sqrt{(c/2m)^2 - (k/m)}$$

and

$$s_1 = -\frac{c}{2m} - \sqrt{(c/2m)^2 - (k/m)}$$

The general solution is

$$x = Ae^{s_1 t} + B\,e^{s_2 t}$$

The constants A and B are determined by the initial conditions $x(0)$ and $\dot{x}(0)$. The single spring–mass system exhibits three types of behavior, overdamped, underdamped and critically damped.

The overdamped state exists when k/m is larger than $(c/2m)^2$. No oscillations exist in this state. The underdamped case is oscillatory and results when $(c/2m)^2$ is larger than k/m. The limiting case between oscillatory and nonoscillatory motion occurs when $(c/2m)^2 = k/m$. When this condition, is met, the system is said to be critically damped.

The next level of complexity results when the system is forced harmonically. The following equation serves as a model for this system:

$$m\ddot{x} + c\dot{x} + kx = F \sin \omega t$$

The solution of this equation is found by first computing the complementary function, which is the solution of the homogeneous equation discussed above, and then the particular solution. The details for computing a particular solution can be found in books on differential equations or mechanical vibrations (Thomson and Dahleh, 1998).

1.2.1 Forced Single-Degree-of-Freedom System

In general, when the oscillatory system is forced in a nonharmonic way, the resulting differential equation cannot be solved in closed form, and numerical methods must be employed to predict the behavior of the system. In this section, we consider two finite difference methods chosen for their simplicity. For a more detailed discussion of numerical methods for ordinary differential equations see Atkinson (1978) and Isaacson (1966).

The spring–mass system that is subjected to general forcing, $F(x, \dot{x}, t)$, can be modeled by the following differential equation:

$$m\ddot{x} + c\dot{x} + kx = f(x, \dot{x}, t)$$

$$x_0 = x(0)$$

$$\dot{x}_0 = \dot{x}(0)$$

where m, c, and k are constants and F is an arbitrary function. The two initial conditions, x_0 and $\dot{x}_0$, are known.

In the first method, the second-order equation is integrated without change in form; in the second method, the second-order equation is rewritten as a system of two first-order equations and then the system of equations is integrated. Both methods approximate the first and second derivatives with the centered difference approximation for the derivatives. Finite difference methods are based on the Taylor expansion. The centered difference method for the first and second derivative results from a combination of the forward and backward Taylor expansion about the point x_i. To get the forward expansion, one writes the Taylor expansion for x_{i+1}. Similarly, the backward expression is obtained from the Taylor expansion about x_{i-1}. These are given by

$$x_{i+1} = x_i + h\dot{x}_i + \frac{h^2}{2}\ddot{x}_i + \frac{h^3}{6}\dddot{x} + \cdots$$

$$x_{i-1} = x_i - h\dot{x}_i + \frac{h^2}{2}\ddot{x}_i - \frac{h^3}{6}\dddot{x} + \cdots$$

where $h = \Delta t$. A second-order approximation is one which matches the Taylor expansion exactly up to and including terms of order h^2; that is, to determine a second-order expansion one neglects terms of order h^3 and higher. A second-order approximation for the first derivative is obtained by subtracting the backward difference from the forward difference. The resulting centered difference approximation is given by

$$\dot{x}_i = \frac{1}{2h}(x_{i+1} - x_{i-1}) - \frac{h^2}{6}\dddot{x} + \cdots$$

Errors result when this expression is truncated after the first term. These errors depend on h^2 and the third derivative of x_i. If the error in the computed derivative is larger than order h^2 it may well arise from the neglected third derivative term.

The centered difference approximation for the second derivative is found by adding the forward and backward difference expansions and ignoring terms of order h^4 and higher. The resulting approximation is

$$\ddot{x}_i = \frac{1}{h^2}(x_{i-1} - 2x_i + x_{i+1}) + \frac{h^2}{12}x^{iv} + \cdots$$

The truncation error that occurs for this expression depends on h^2 and the fourth derivative of x_i. Both the first and second derivative centered difference approximations are order h^2. They can be used together to create a second-order approximation to a second-order ordinary differential equation such as that describing the spring–mass system.

1.2.1.1   Centered Difference Approximation

After substituting these two centered difference approximations into the differential equation and rearranging terms, one gets the following recurrence relation for the single spring–mass system:

$$x_{i+1} = h^2 f(x_i, t_i) - (fh^2 - 2m)x_i - (m - ch/2)x_{i-1}$$

with the initial conditions

$$x_0 = x(0)$$
$$\dot{x}_0 = \dot{x}(0)$$

The recurrence relation should be used to compute all values of x from the initial condition. By letting $i = 1$ in the recurrence relation, we get

$$x_2 = h^2 f(x_1, t_1) - (kh^2 - 2m)x_1 - (m - ch/2)x_0$$

In order to compute x_2, both x_0 and x_1 are needed. The initial conditions provide x_0. However, to start the calculation, we need an additional equation for x_1. This equation is derived by substituting $i = 0$ into the Taylor expansion for x_{i+1}, using the initial conditions and ignoring terms of order h^2 and higher. Since the centered difference approximation is of order h^2, it is consistent to compute x_1 with an error of order h^2. For the point $i = 0$, the forward difference is given by

$$x_1 = x_0 + h\dot{x}_0$$

This equation allows one to find x_1 in terms of the two initial conditions, after which the recurrence relation can be used to find all subsequent discrete values of x.

1.2.1.2   Pseudocode for Centered Difference Approximation

The following is an example of the MATLAB® routine for the solution to the centered difference approximation of the general single-DoF spring–mass equations. The function f needs to be specified in a separate function file:

```
%initial conditions (index has to start at 1)
x(1) = x₀
xd = x'₀
%specify a time step
h = H
%specify the constants m, k, c.
m = M
k = K
c = C
%compute x(2) from the Taylor expansion
x(2) = x(1) + h * xd
%specify the total number of steps
TEND = tfinal
```

%compute the solution for all remaining times using the recurrence relation
$t(1) = h$
$t(2) = 2 * h$
for $I = 3$:TEND
$X_i = h^2 * f(x_{i-1}, t_{i-1}) - (k * h^2 - 2 * m) * x_{i-1} - (m - c * h/2) * x_{i-2}$
$T(i) = t(I - 1) + h$
end

The method just presented has ignored terms of order h^2 and higher. This is known as the truncation error. The calculation will contain other errors such as round-off error due to the loss of significant figures. This loss is related to both the machine and the language used for the calculations. The round-off errors are also related to the time increment $h = \Delta t$ in a complicated way that is beyond the scope of this chapter. Better accuracy can be obtained by choosing a smaller Δt. However, the smaller the Δt, the larger the number of computations needed to reach the solution in a fixed time T. The increased number of computations affects both the total time of the calculation and the overall accuracy.

1.2.1.3 Runge–Kutta Methods

The centered difference approximation is not a self-starting method. In other words, the calculation that determines x_1 from the initial conditions does not come from the discretized equation, but rather from the Taylor expansion directly. An alternate way to solve this equation is to use what is known as a Runge–Kutta method. All of these methods approximate the differential equation with Taylor series expansions. The advantage of this class of methods is that they are self-starting. The disadvantage is that they only work on first-order equations (or systems of first-order equations). Before the Runge–Kutta method is discussed in detail, the second-order spring–mass equation has to be written as a system of two first-order equations. The equation for the single-DoF system, subjected to arbitrary forcing f is

$$m\ddot{x} + c\dot{x} + kx = f(x, \dot{x}, t)$$

It can be rewritten as the following system of first-order equations

$$\dot{x} = u$$

$$\dot{u} = \ddot{x} = \frac{1}{m}[f(x, u, t) - cu + kx]$$

These equations are coupled and need to be solved together. Any Runge–Kutta method for the system of equations requires a Taylor series expansion of x and u about x_i and u_i. The Taylor expansion is given by

$$x_{i+1} = x_i + \dot{x}_i h + \ddot{x}_i \frac{h^2}{2} + \cdots$$

$$u_{i+1} = u_i + \dot{u}_i h + \ddot{u}_i \frac{h^2}{2} + \cdots$$

As above, the time increment will be denoted by $h = \Delta t$. The first-order Runge–Kutta method (also known as Euler's method) is obtained by retaining terms of first-order and lower, i.e., the Euler approximation is given by

$$x_{i+1} = x_i + \dot{x}_i h - o(h)$$

$$u_{i+1} = u_i + \dot{u}_i h - o(h)$$

The truncation error for Euler's method is order h. The error depends linearly on h.

Matching more terms in the Taylor expansion generates higher order Runge–Kutta methods. The most commonly used Runge–Kutta method is the fourth-order method that matches the Taylor expansion up to terms of order h^4. This is a significant reduction in the error. For a derivation of this method see Cheney and Kincaid (1999).

TABLE 1.1 Fourth-Order Runge–Kutta Method for Spring–Mass Equation

t	x	$\dot{X} = u$	$\ddot{X} = G$
$T_1 = t_i$	$X_1 = x_i$	$U_1 = u_i$	$G_1 = G(T_1, X_1, U_1)$
$T_2 = t_i + h/2$	$X_2 = x_i + U_1 h/2$	$U_2 = u_i + G_1 h/2$	$G_2 = G(T_2, X_2, U_2)$
$T_3 = t_i + h/2$	$X_3 = x_i + U_2 h/2$	$U_3 = u_i + G_2 h/2$	$G_3 = G(T_3, X_3, U_3)$
$T_4 = t_i + h$	$X_4 = x_i + U_3 h$	$U_4 = u_i + G_3 h$	$G_4 = G(T_4, X_4, U_4)$

Source: Thomson and Dahleh 1998. *Theory of Vibration Applications*, 5th ed. With permission.

For the system of first-order equations given above, the fourth-order Runge–Kutta method requires four values of t, x, u, and, G where $G = 1/m[f(x, u, t) - cu + kx]$. It can be computed for each point i as follows in Table 1.1.

Combining these quantities in the following method gives the fourth-order Runge–Kutta method:

$$x_{i+1} = x_i + h/6(U_1 + 2U_2 + 2U_3 + U_4)$$

$$u_{i+1} = u_i + h/6(G_1 + 2G_2 + 2G_3 + G_4)$$

where it is recognized that the four values of U divided by six represent the average slope dx/dt and the four values of G divided by six result in an average of du/dt. A way to check the accuracy of a Runge–Kutta method is to Taylor expand G_1, G_2, G_3, and G_4 and collect like terms. One will find that the above combination produces an expansion which is exact up to order h^4.

1.2.1.4 Pseudocode for the Fourth-Order Runge–Kutta Method

For simplicity, the code is given for the single first-order equation

$\dot{x} = G(t, x)$ with the initial data $x(0) = x_0$.

The function G should be specified in a function file.

```
%initial conditions (index has to start at 1)
x(1) = x0;
%time step needs to be specified
h = H;
h2 = 2 * H;
%TEND is the total number of time steps
TEND = Tfinal;
T = h;
for I = 1:TEND
G1 = H * G(t, x);
G2 = H * G(t + h2, x = 0.5 * G1);
G3 = H * G(t + h2, x = 0.5 * G2);
G4 = h * G(t + h, x + G3);
X = x + (G1 + 2 * G2 + 2 * G3 + G4)/6.0;
t = (I + 1) * h;
end
```

1.2.1.5 Example

Solve numerically the differential equation

$$4\ddot{x} + 2000x = F(t)$$

with the initial conditions

$$x_0 = \dot{x}_0 = 0$$

The forcing is as shown in Figure 1.2.

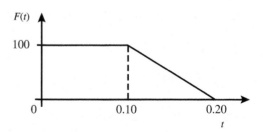

FIGURE 1.2 The forcing function for Example 1.2.1.5. (*Source*: Thomson and Dahleh 1998. *Theory of Vibration Applications*, 5th ed. With permission.)

1.2.1.5.1 Centered Difference

Using the centered difference approximation for the second derivative, one gets the following discrete equation

$$x_{i+1} = h^2/4F(t) - 500h^2 x_i - x_{i-1} + 2x_i$$

This equation is valid for $i \geq 1$. From the initial conditions $x_0 = \dot{x}_0 = 0$, x_1 is computed by the following $x_1 = x_0 + \dot{x}_0 = 0 + 0$. A time step of $h = 0.02$ sec has been used in the calculation. The numerical solution as compared with the exact solution is given in Table 1.2. Table 1.3 contains the absolute errors for the two discrete solutions.

1.2.1.5.2 Runge–Kutta

In order to use the Runge–Kutta method, the second-order equation needs to be written as a system of first-order equations.

Let $u = \dot{x}$; then

$$\dot{u} = G(x, t) = .25 * F(t) - 500x$$

where $x(0) = 0$ and $u(0) = 0$.

This is now in a form which can be directly input into a fourth-order Runge–Kutta solver.

The Runge–Kutta method is more accurate then the centered difference approximation. This can be seen in Table 1.3, which gives the absolute value of the difference between the exact solution and the computed solutions. This is known as the absolute error. Moreover, the Runge–Kutta method is self-starting and can be used for a single variable or a system of variables as in the above example. The price that is paid is that the fourth-order Runge–Kutta method requires four function evaluations for the first derivative for each time step. This is offset by the fact that this method has higher accuracy and has a large stability region so one can take larger time steps.

TABLE 1.2 Solution to Example 1.2.1.5

Time t	Exact Solution	Central Difference	Runge–Kutta
0	0	0	0
0.02	0.00492	0.00500	0.00492
0.04	0.01870	0.01900	0.01869
0.06	0.03864	0.03920	0.03862
0.08	0.06082	0.06159	0.06076
0.10	0.08086	0.08167	0.08083
0.12	0.09451	0.09541	0.09447
0.14	0.09743	0.09807	0.09741
0.16	0.08710	0.08712	0.08709
0.18	0.06356	0.06274	0.06359
0.20	0.02949	0.02782	0.02956
0.22	-0.01005	-0.01267	-0.00955
0.24	-0.04761	-0.05063	-0.04750
0.26	-0.07581	-0.07846	-0.07571
0.28	-0.08910	-0.09059	-0.08903
0.30	-0.08486	-0.08461	-0.08485
0.32	-0.06393	-0.06171	-0.06400
0.34	-0.03043	-0.02646	-0.03056
0.36	0.00906	0.01407	0.00887
0.38	0.04677	0.05180	0.04656
0.40	0.07528	0.07916	0.07509
0.42	0.08898	0.09069	0.08886
0.44	0.08518	0.08409	0.08516
0.46	0.06436	0.06066	0.06423
0.48	0.03136	0.02511	0.03157

TABLE 1.3 Absolute Error for Centered Difference and for the Fourth-Order Runge–Kutta

Time t	Error for Central Difference	Error for Runge–Kutta All Times 1.0×10^{-3}
0	0	0
0.02	0.0001	0.0
0.04	0.0003	0.0100
0.06	0.0006	0.0180
0.08	0.0008	0.0600
0.10	0.0008	0.0300
0.12	0.0009	0.0400
0.14	0.0006	0.0200
0.16	0.0000	0.0100
0.18	0.0008	0.0300
0.20	0.0017	0.0700
0.22	0.0026	0.0500
0.24	0.0030	0.1100
0.26	0.0026	0.1000
0.28	0.0015	0.0700
0.30	0.0003	0.0100
0.32	0.0022	0.0700
0.34	0.0040	0.1300
0.36	0.0050	0.1900
0.38	0.0050	0.2100
0.40	0.0039	0.1900
0.42	0.0017	0.1200
0.44	0.0011	0.0200
0.46	0.0037	0.1300
0.48	0.0062	0.2100

Stability is a measure of how quickly errors in the computed solution grow or decay. There are very few numerical methods that are stable for all choices of time step. For most methods, there is a range of time steps which produce a stable method. The fourth-order Runge–Kutta method is stable for larger values of the time step than are the lower-order Runge–Kutta methods. A numerical method will not converge to a solution if the time step does not produce a stable method. Often, the stability criterion places a stricter limitation on the time step than accuracy does. For a more complete discussion of stability, see Strang (1986). In Table 1.3, the absolute error for both methods grows but it does not grow exponentially. Controlled error growth is the signature of a stable time step.

1.2.2 Summary of Single-Degree-of-Freedom System

- Unforced equation of motion $m\ddot{x} + c\dot{x} + kx = 0$.
- Forced equation of motion $m\ddot{x} + c\dot{x} + kx = f(x, \dot{x}, t)$.
- Centered difference approximation for the first derivative $\dot{x}_i = \dfrac{1}{2h}(x_{i+1} - x_{i-1})$.
- Centered difference approximation for the second derivative $\ddot{x}_i = \dfrac{1}{h^2}(x_{i-1} - 2x_i + x_{i+1})$.
- Fourth-order Runge–Kutta method.

1.3 Systems with Two or More Degrees of Freedom

A system that requires more than one coordinate to describe its motion is a multi-DoF system. These systems differ from single-DoF systems in that n DoF are described by n simultaneous differential equations and have n natural frequencies. When these systems are written in matrix notation, they look just like the single-DoF system. The equations of motion of a viscously damped multi-DoF system can be written as follows:

$$[\mathbf{M}]\ddot{x} + [\mathbf{C}]\dot{x} + [\mathbf{K}]\mathbf{x} = \mathbf{f}$$

where [**M**], [**C**], and [**K**] are the mass, damping, and stiffness matrices, respectively; **x** is the displacement vector; and **f** is the force vector. Both the central difference method and Runge–Kutta can be applied to the matrix equation. The method follows exactly that given above where the scalar quantities are replaced by the matrix quantities.

1.3.1 Example

As an example, consider the two-DoF system shown in Figure 1.3.

In this example, $k_1 = k_2 = 36$ KN/m, $m_1 = 100$ kg, $m_2 = 25$ kg, and $f = 4000$ N for $t > 0$ and 0 for $t \leq 0$. The initial conditions are all zero, i.e., $x_1 = \dot{x}_1 = x_2 = \dot{x}_2 = 0$. The equation of motion for this system is

$$100\ddot{x}_1 + 36{,}000x_1 - 36{,}000(x_2 - x_1) = 0$$

$$25\ddot{x}_2 + 36{,}000(x_2 - x_1) = f$$

This can be written in matrix notation as

$$\begin{bmatrix} 100 & 0 \\ 0 & 25 \end{bmatrix} \ddot{x} + 36{,}000 \begin{bmatrix} 2 & -1 \\ -1 & 1 \end{bmatrix} = \mathbf{f}$$

where $\mathbf{x} = (x_1, x_2)^t$ and $\mathbf{f} = (0, f)^t$.

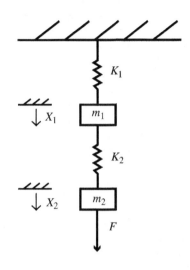

FIGURE 1.3 Two-DoF system. (*Source:* Thomson and Dahleh 1998. *Theory of Vibration Applications*, 5th ed. With permission.)

1.3.1.1 Centered Difference

Using the centered difference approximation for the second derivatives, one obtains the following two recurrence relations

$$x_1^{i+1} = (-720x_1^i + 360x_2^i)\Delta t^2 + 2x_1^i - x_1^{i-1}$$

$$x_2^{i+1} = (1440(x_1^i - x_2^i) + 160)\Delta t^2 + 2x_2^i - x_2^{i-1}$$

These two equations are only valid for $i \geq 3$. From the single-DoF system, we know that this method is not self-starting. In order to compute x_1^2 and x_2^2, one needs to use an additional equation, which one obtains from the Taylor expansion. First one needs to compute the initial acceleration for the system. This is obtained from the differential equation and the initial data. For this problem, the initial acceleration is given by $\ddot{x}_1 = 0$ and $\ddot{x}_2 = 160$. Next, values for x_1^{-1} and x_2^{-1} are computed from the Taylor expansion

$$x_1^{-1} = x_1^0 - \Delta t \dot{x}_1 + \frac{\Delta t^2}{2} \ddot{x}_1 = 0$$

$$x_2^{-1} = 80\Delta t^2$$

Now, the recurrence relations can be used to compute the rest of the terms. An example of the MATLAB code for this calculation is the following.

1.3.1.2 Pseudocode for Example 1.3.1

The following is a MATLAB m file for Example 1.3.1.

```
clear
deltat = 0.01;
deltsq = deltat*deltat;
x2(1) = 0;
x1(1) = 0;
x2(2) = 160/2*deltsq;
x1(2) = (60*x2(2)*deltsq)/(1 + 120*deltsq);
x2ddot(2) = 1440*(x1(2) - x2(2)) + 160;
x1ddot(2) = -720*x1(2) + 360*x2(2);
for i = 3:51
    x2(i) = x2ddot(i - 1)*deltsq + 2*x2(i - 1) - x2(i - 2);
    x1(i) = x1ddot(i - 1)*deltsq + 2*x1(i - 1) - x1(i - 2);
    x2ddot(i) = 1440*(x1(i) - x2(i)) + 160;
    x1ddot(i) = -720*x1(i) + 360*x2(i);
end
```

Figure 1.4 gives the displacement of $x1$ and $x2$ over time for Example 1.3.1.

1.3.1.3 Runge–Kutta

Alternately, the system of second-order equations can be integrated using the fourth-order Runge–Kutta method. In order to use Runge–Kutta, the second-order system has to be written as a system of first-order equations. The following second-order system

$$\ddot{x}_1 = -720x_1 + 360x_2$$

$$\ddot{x}_2 = 1440(x_1 - x_2) + 160$$

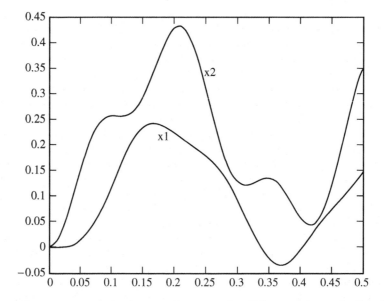

FIGURE 1.4 Displacement versus time. (*Source*: Thomson and Dahleh 1998. *Theory of Vibration Applications*, 5th ed. With permission.)

becomes the following first-order system

$$\dot{x}_1 = x_3$$

$$\dot{x}_2 = x_4$$

$$\dot{x}_3 = -720x_1 + 360x_2$$

$$\dot{x}_4 = 1400(x_1 - x_2) + 160$$

Two new variables have been introduced.

1.3.1.4 Integrating Ordinary Differential Equations Using MATLAB

There are several ODE solvers in MATLAB, especially in version six. The discussion here will be limited to the least complicated solvers. These are *ode23* and *ode45*, which are implementations of a second- and third-order, and a fourth- and fifth-order solver, respectively. Like all Runge–Kutta methods, these solvers work on first-order equations. If the equation is of higher order, it needs to be converted to a system of first-order equations. Then the equation should be written in vector form, i.e., $\mathbf{x}' = \mathbf{f}(x, t)$. The user must write a MATLAB function routine which computes the values of $\mathbf{f} = (f_1, f_2, \ldots f_n)$ given the values of $(\mathbf{x}, t)$. Once this function files exists, it can be used as input in either ode23 or ode45. The function call details are given in the on-line help facility in MATLAB. See the Appendix for an introduction to MATLAB.

1.3.2 Summary of Two-Degree-of-Freedom System

1. Equation of motion: $[\mathbf{M}]\ddot{x} + [\mathbf{C}]\dot{x} + [\mathbf{K}]\mathbf{x} = \mathbf{f}$.
2. Centered difference for systems of equations.
3. Runge–Kutta methods.
4. MATLAB commands for solution of ordinary differential equations — ode23 and ode45.

1.4 Finite Difference Method for a Continuous System

1.4.1 Bar

A thin uniform rod is the first example of a continuous system that will be examined. The equation of motion for such a system is given by

$$\partial^2 u/\partial x^2 = 1/c^2 \partial^2 u/\partial t^2$$

where $c = \sqrt{E/\rho}$. E is the modulus of elasticity and ρ is the density of the rod. A complete derivation of this equation can be found in many books (see Thomson and Dahleh 1998). Separation of variables can be used to find a solution of the form $u(x, t) = U(x)G(t)$. This substitution gives the equation.

$$\frac{d^2 U}{dx^2} G(t) = \frac{1}{c^2} U(x) \frac{d^2 G(t)}{dt^2}$$

Upon rearrangement, this becomes

$$\frac{1}{U(x)} \frac{d^2 U}{dx^2} = \frac{1}{c^2 G(t)} \frac{d^2 G(t)}{dt^2}$$

The left-hand side of the equation is independent of t and the right-hand side is independent of x. Therefore, each side must equal a constant. Let this constant be denoted by α and $\alpha = -(\omega/c)^2$. There are now two differential equations, the displacement equation

$$d^2 U/dx^2 + \alpha^2 U = 0$$

and the temporal equation

$$d^2G/dt^2 + \omega^2 G = 0$$

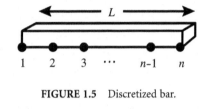

The displacement equation can be solved using
a finite difference approximation. The finite
difference approximation requires dividing the

FIGURE 1.5   Discretized bar.

bar of length L into $n - 1$ pieces each of length
$h = L/(n - 1)$. The finite difference mesh consists of n points, $1, 2, 3, \ldots, n$. The first node (labeled 1)
is the left-hand endpoint of the bar, and the last node, labeled n, is the right-hand endpoint of the bar
(see Figure 1.5). The value of U at each point i is denoted by U_i. Using the centered difference
approximation for the second derivative, the discretized version of the displacement equation becomes

$$(U_{i+1} - 2U_i + U_{i-1})/h^2 + \alpha U_i = 0$$

Collecting like terms, the equation becomes

$$U_{i+1} - (2 - \lambda)U_i + U_{i-1} = 0$$

where $\lambda = h^2 \alpha^2$. As is typical for the finite difference approximation, this equation holds at the
interior mesh points $(i = 2, 3, \ldots, n - 1)$ and not at the endpoints. To see that this equation does not
hold at the endpoints, substitute $i = 1$ into the recurrence relationship. The resulting equation is

$$U_2 - (2 - \lambda)U_1 + U_0 = 0$$

The problem is that there is no value U_0. Recall that U_1 represents the displacement of the left-
hand endpoint. Since U_0 represents a point even further to the left, it is not on the bar. Similarly, the
recurrence relation does not work for $i = n$. When the relationship is evaluated for $i = n$, the value
U_{n+1} is needed and it does not exist. The displacement equation evaluated for $i = n$ is

$$U_{n+1} - (2 - \lambda)U_n + U_{n-1} = 0$$

When the relationship is evaluated for $(i = 2, 3, \ldots, n - 1)$, the resulting matrix problem is tri-
diagonal. This is characteristic of the centered difference approximation. It is given by the following
matrix:

$$\begin{bmatrix} -1 & 2-\lambda & -1 & 0 & \ldots & 0 \\ 0 & -1 & 2-\lambda & -1 & 0 & \ldots & 0 \\ 0 & 0 & -1 & 2-\lambda & -1 & \ldots & 0 \\ & & & & & & \\ & & 0 & -1 & 2-\lambda & -1 \end{bmatrix} \begin{bmatrix} U_1 \\ U_2 \\ U_3 \\ \\ U_n \end{bmatrix} = \begin{bmatrix} 0 \\ 0 \\ 0 \\ \\ 0 \end{bmatrix}$$

This matrix problem does not have a unique solution since there are n unknowns and there are
only $n - 2$ equations. The boundary conditions for the bar provide the needed two additional
constraints. For a bar that is fixed at both ends, the deflection is zero at the ends ($x = 0$ and $x = L$).
Under these conditions, $U_1 = U_n = 0$. If one of the ends is free then the stress is zero at that end,
which means that $dU/dx = 0$. For the purposes of illustration, let us assume that the end at $x = 0$ is
fixed and the other end is free. This means that $U_1 = 0$ and $dU/dx|n = 0$. If we use the centered
difference approximation for the derivative, we get

$$dU/dx|n = (U_{n+1} - U_{n-1})/2h = 0$$

which means that $U_{n+1} = U_{n-1}$.

U_{n+1} is the displacement of a point off the bar. We never compute U_{n+1}; however, as one can see from a previous calculation, this value is used when the recurrence relationship is evaluated for $i = n$, $U_{n+1} - (2 - \lambda)U_n + U_{n-1} = 0$. Under the assumption that the stress is zero at the right-hand endpoint, the remaining constraint becomes

$$(2 - \lambda)U_n - 2U_{n-1} = 0$$

For a bar that has one fixed left-hand endpoint and a free right-hand end, the matrix problem is the following:

$$\begin{bmatrix} 2 - \lambda & -1 & 0 & 0 & \cdots & 0 \\ -1 & 2 - \lambda & -1 & 0 & \cdots & 0 \\ 0 & -1 & 2 - \lambda & -1 & \cdots & 0 \\ & & & & & \\ 0 & & & -1 & 2 - \lambda & -1 \\ & & & & -2 & 2 - \lambda \end{bmatrix} \begin{bmatrix} U_2 \\ U_3 \\ U_4 \\ \\ U_{n-1} \\ U_n \end{bmatrix} = \begin{bmatrix} 0 \\ 0 \\ 0 \\ \\ 0 \\ 0 \end{bmatrix}$$

This matrix problem has a unique solution because there are $n - 1$ linearly independent equations and there are $n - 1$ unknowns.

1.4.1.1 Example

Consider a bar of unit length that is divided into four equal pieces. The left-hand end is fixed so that $U_1 = 0$, and the right-hand end is free. The eigenvalue problem for this system is the following:

$$\begin{bmatrix} 2 - \lambda & -1 & 0 & 0 \\ -1 & 2 - \lambda & -1 & 0 \\ 0 & -1 & 2 - \lambda & -1 \\ 0 & 0 & -2 & 2 - \lambda \end{bmatrix} \begin{bmatrix} U_2 \\ U_3 \\ U_4 \\ U_5 \end{bmatrix} = \begin{bmatrix} 0 \\ 0 \\ 0 \\ 0 \end{bmatrix}$$

The eigenvalues, λ, are $\lambda = [0.1522, 1.2346, 2.7654, 3.8478]$. The natural frequencies can be recovered from the eigenvalues since $\lambda = h^2 \omega/c$ and $c = \sqrt{E/\rho}$ is determined by the material properties of the bar.

1.4.2 Beam

A second example of finite difference approximation is given by looking at a centered difference approximation to compute the transverse vibration of a uniform beam. We will consider a special case of the Euler equation for the beam. The following fourth-order equation results when the flexural rigidity, (EI), is constant:

$$d^4 W/dx^4 - \beta^4 W = 0$$

where β contains all of the material properties of the beam (i.e., $\beta^4 = \rho A \omega^2/EI$).

The centered difference formula for the fourth derivative is given by

$$f^{iv} = 1/h^4 (f_{i+2} - 4f_{i+1} + 6f_i - 4f_{i-1} + f_{i-2})$$

In order to see that this expansion is in fact second-order, combine the Taylor expansions for f_{i+2}, f_{i+1}, f_i, f_{i-1}, and f_{i-2}, and notice that you obtain an approximation for the fourth derivative which exactly matches the combined Taylor approximations up to order h^2. Using this approximation, the transverse beam equation at the interior mesh point becomes

$$W_{i+2} - 4W_{I+1} + (6 - \lambda)W_i - 4W_{i-1} + W_{i-2} = 0$$

where $\lambda = h^4\beta^4$. This equation is valid for $i = 3, n - 2$. As before, the boundary conditions have to be used in order to have sufficent conditions to determine all n values for the deflection, W. There will be two additional constraints placed at each end. In this case, all of the types of boundary condition require fictitious points. In other words, they require that the beam be extended beyond the physical boundaries to include W_{-1}, W_0, W_{n+1}, and W_{n+2}. Let us consider three types of boundary conditions: fixed end, free end, and simply-supported end. For simplicity, we will derive all of the conditions for the right-hand endpoint $x = L$.

Fixed end: When the end of the beam is fixed, both the deflection $W = 0$ and $dW/dx = 0$. As above, the centered difference approximation for the first derivative requires the introduction of a fictitious point. These conditions translate into

$$W_n = 0 \text{ and } dW/dx|_n = 1\backslash h^2(W_{n+1} - W_{n-1}) = 0$$

These constraints reduce to the following:

$$W_{n+1} = W_{n-1}$$

Free end: The bending moment and the shear force are zero at the free end. These are given by a second and third derivative, respectively. Assuming the free end is located at N, the two fictitious points introduced are at $n + 1$ and $n + 2$. The discrete version of the two constraints is:

$$d^2W/dx^2|_n = 1/h^2(W_{n-1} - 2W_n + W_{n+1}) = 0$$

$$d^3W/dx^3|_n = 1/(2h^3)(W_{n+2} - 2W_{n+1} + 2W_{n-1} - W_{n-2}) = 0$$

Simply-supported end: The boundary conditions of this type require that the deflection and the moment are zero. When these are applied to the right-hand endpoint

$$W_n = 0$$
$$d^2W/dx^2 = 1/h^2(W_{n+1} - 2W_n + W_{n-1}) = 0$$

Combining these two equations, one sees that the value of W at the fictitious point is equal to the value on the beam, i.e., $W_{n+1} = W_{n-1}$.

All three of these boundary condition types allow one to replace the fictitious values in the recurrence relations with values on the beam. The procedures followed are very similar to those outlined for the longitudinal vibration of a beam.

1.4.3   Summary of Finite Difference Methods for a Continuous System

1. Equation of motion for the bar: $\partial^2 u/\partial x^2 = 1/c^2 \partial^2 u/\partial t^2$.
2. Displacement equation for the bar: $d^2U/dx^2 + \alpha^2 U = 0$.
3. Finite difference approximation: $U_{i+1} - (2 - \lambda)U_i + U_{i-1} = 0$.
4. Equation of motion for the beam: $d^4W/dx^4 - \beta^4 W = 0$.
5. Centered difference approximation for the fourth derivative: $f^{iv} = 1/h^4(f_{i+2} - 4f_{i+1} + 6f_i - 4f_{i-1} + f_{i-2})$.
6. Finite difference approximation for the beam equation: $W_{i+2} - 4W_{i+1} + (6 - \lambda)W_i - 4W_{i-1} + W_{i-2} = 0$.

1.5   Matrix Methods

In the preceding section, the equations of motion are solved for the system. Often, one is not interested in the complete solution. Rather, one is interested in the natural frequencies and the normal modes. When the number of DoF is very high, only the lowest natural frequency needs to be computed. Several methods to find some, but not all, of the eigenvalues of the system will be presented.

1.5.1 Example: Three-Degree-of-Freedom System

A three-DoF system will be used as an example. The larger DoF systems work exactly in the same manner but produce larger calculations.

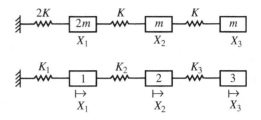

Let us consider the following spring–mass system in Figure 1.6.

The equation of motion for this system is

$$M \begin{bmatrix} 2 & 0 & 0 \\ 0 & 1 & 0 \\ 0 & 0 & 1 \end{bmatrix} \begin{bmatrix} \ddot{x}_1 \\ \ddot{x}_2 \\ \ddot{x}_3 \end{bmatrix} + K \begin{bmatrix} 3 & -1 & 0 \\ -1 & 2 & -1 \\ 0 & -1 & 1 \end{bmatrix} \begin{bmatrix} x_1 \\ x_2 \\ x_3 \end{bmatrix}$$

$$= \begin{bmatrix} 0 \\ 0 \\ 0 \end{bmatrix}$$

FIGURE 1.6 Three spring–mass system. (*Source:* Thomson and Dahleh 1998. *Theory of Vibration Applications*, 5th ed. With permission.)

Assuming a harmonic solution of the form $x = A \sin(\omega t)$ gives the following eigenvalue problem:

$$\left(-\lambda \begin{bmatrix} 2 & 0 & 0 \\ 0 & 1 & 0 \\ 0 & 0 & 1 \end{bmatrix} + k \begin{bmatrix} 3 & -1 & 0 \\ -1 & 2 & -1 \\ 0 & -1 & 1 \end{bmatrix} \right) \begin{bmatrix} x_1 \\ x_2 \\ x_3 \end{bmatrix} = \begin{bmatrix} 0 \\ 0 \\ 0 \end{bmatrix}$$

where $\lambda = \omega^2 m/k$. The eigenvalues are found by setting the characteristic equation of the determinant equal to zero. This gives the following polynomial:

$$\lambda^3 - 4.5\lambda^2 + 5\lambda - 1 = 0$$

It is a simple matter to graph this polynomial and see that there are three real roots to this equation.

1.5.2 Bisection Method

One method for calculating the roots of a nonlinear function is called the bisection method. This method finds the zeros of nonlinear functions by bracketing the zero in an interval [a, b]. The interval is chosen so that $f(a)$ and $f(b)$ are of opposite sign. If f is a continuous function and it is positive at one endpoint, say $f(a) > 0$, and it is negative at the other endpoint, $f(b) < 0$, then it has had to go through zero at some point in the interval. It is possible that it has gone through zero more then once in the interval. From Figure 1.7, one can see that one root is between [0, 0.5]; another is between [1, 1.5]; and a third root is in [2.5, 3]. In order to find all three roots, the bisection method has to be used three times; one for each interval. As an example, the MATLAB code for the first root is given below.

1.5.2.1 MATLAB Code for the Bisection Method

```
clear
%set an acceptable tolerance for the root
tol = 10e − 4
%endpoints of the interval
a = 0
b = 0.5
for i = 1:20
c = 0.5*(a + b)
if abs(f(c)) < tol, break, end
```

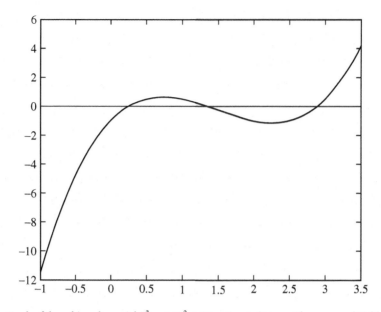

FIGURE 1.7 Graph of the cubic polynomial $\lambda^3 - 4.5\lambda^2 + 5\lambda - 1 = 0$. (*Source*: Thomson and Dahleh 1998. *Theory of Vibration Applications*, 5th ed. With permission.)

```
if f(a)*f(c) < 0
    b = c
else
    a = c
end
end
```

The root 0.2554 is found after ten iterations. As long as the initial interval is chosen so that the function has different signs at the endpoints, this method will always converge. There are nonlinear root finding methods, such as Newton's method, which when they converge do so more quickly. However, they may or may not converge.

1.5.2.2 MATLAB Function for Finding the Roots of a Polynomial

MATLAB has a built in root-finding method which requires the creation of a vector **c** that contains the coefficients of the polynomial in descending order. The MATLAB command *roots* (*c*), produces the roots of the polynomial. For our example, $\mathbf{c} = [1, -4.5, 5, -1]$. The roots of this equation are given by $\lambda = 2.8892, 1.3554, 0.2554$.

1.5.2.3 Mode Shapes

The mode shapes are determined by substituting each eigenvalue into the matrix problem and computing the corresponding eigenvector. As an example of how to compute the eigenvector for a given eigenvalue, we will compute the eigenvector associated with $\lambda = 2.8892$. The first step is to substitute this value of λ into the eigenvalue equation. This gives the following problem:

$$\begin{bmatrix} 2.489 & -1 & 0 \\ -1 & 1.745 & -1 \\ 0 & -1 & 0.744 \end{bmatrix} \begin{bmatrix} x_1 \\ x_2 \\ x_3 \end{bmatrix} = \begin{bmatrix} 0 \\ 0 \\ 0 \end{bmatrix}$$

The solution vector found by Gaussian elimination is given by

$$\mathbf{x} = (x_1, x_2, x_3)^t = (0.2992, 0.7446, 1.0)^t$$

1.5.3 Directly Calculating the Eigenvalues and Eigenvectors from the Matrix Equation

Ideally one would like to use a computer program such as MATLAB to compute both the eigenvalues and the eigenvectors directly from the matrix equation. In order to do this, the matrix equation needs to be rewritten slightly. In general, the matrix equation for the normal mode vibration is given by

$$[-\lambda \mathbf{M} + \mathbf{K}]x = 0$$

where **M** and **K** are the mass and stiffness matrices, respectively, both are square symmetric matrices, and λ is the eigenvalue related to the natural frequency by $\lambda = n\omega^2$. Premultiplying the preceding equation by $\mathbf{M}^{-1}$, we have another form of the equation:

$$[-\lambda I + \mathbf{A}]x = 0$$

where $\mathbf{A} = \mathbf{M}^{-1}\mathbf{K}$ and **A** is called the dynamic matrix. In general, **A** is not symmetric. In order to use MATLAB to compute the eigenvectors and eigenvalues, the matrix must be symmetric. There is a standard transformation of coordinates that converts the matrix problem into a standard eigenvalue problem which can be solved by a computer. Assume that we have a transformation of coordinates which has the following form:

$$x = U^{-1}$$

When this transformation is substituted into the equation, we get

$$[-\lambda \mathbf{M}U^{-1} + \mathbf{K}U^{-1}]y = 0$$

Premultiplying the equation by $\mathbf{U}^{-t}$ gives

$$[-\lambda U^{-t}\mathbf{M}U^{-1} + U^{-t}\mathbf{K}U^{-1}]y = 0$$

From this equation, one can see that if either **M** or **K** equals $\mathbf{U}^t\mathbf{U}$, then the preceding equation is reduced to standard form, at which time it is possible to compute both the eigenvalues and eigenvectors directly. To see how this works, let us assume that $\mathbf{M} = \mathbf{U}^t\mathbf{U}$. The equation becomes

$$[-\lambda I + \mathbf{U}^{-t}\mathbf{K}U^{-1}]y = 0$$

where $\lambda = \omega^2$. This equation is in standard form since $\mathbf{U}^{-t}\mathbf{K}U^{-1}$ is symmetric.

1.5.3.1 Example

To illustrate the use of the dynamic matrix and the standard computer form, we can use MATLAB to calculate the eigenvalues and eigenvectors for Example 1.5.1. The first step is to convert the problem into standard form. For this example, the dynamic matrix is given by

$$A = \begin{bmatrix} 1.5 & -0.5 & 0 \\ -1.0 & 2.0 & -1.0 \\ 0 & -1.0 & 1.0 \end{bmatrix}$$

We can compute both the eigenvalues and the eigenvectors in MATLAB using the command $[U, D] = \text{eig}(A)$. The result of this command is two matrices, **U** and **D**. Matrix **U** contains the eigenvectors as column vectors and **D** is a diagonal matrix which has the eigenvalues on the diagonal. Continuing with our example, we get the following two matrices:

$$\mathbf{U} = \begin{bmatrix} 0.7569 & 0.3031 & 0.2333 \\ 0.2189 & -0.8422 & 0.5808 \\ -0.6158 & 0.4458 & 0.7799 \end{bmatrix}$$

and

$$\mathbf{D} = \begin{bmatrix} 1.3554 & 0 & 0 \\ 0 & 2.8892 & 0 \\ 0 & 0 & 0.2544 \end{bmatrix}$$

In order to use this transformation then, one has to be able to write $\mathbf{M}$ or $\mathbf{K}$ as $\mathbf{U}^t\mathbf{U}$. This is known as the Cholesky decomposition. In MATLAB, if one has a positive definite matrix $\mathbf{M}$, the matrix can be Cholesky-decomposed with the built in function *chol*. The command $U = chol(M)$ produces an upper triangular matrix $\mathbf{U}$ such that $\mathbf{U}^t\mathbf{U} = \mathbf{M}$.

Cholesky decomposition is a special case of LU factorization where L is a lower triangular matrix and U is an upper triangular matrix. The computational procedure begins by writing the algebraic equations that result from the following calculation:

$$\begin{bmatrix} u_{11} & 0 & 0 \\ u_{12} & u_{22} & 0 \\ u_{13} & u_{23} & u_{33} \end{bmatrix} \begin{bmatrix} u_{11} & u_{12} & u_{13} \\ 0 & u_{22} & u_{23} \\ 0 & 0 & u_{33} \end{bmatrix} = \begin{bmatrix} m_{11} & m_{12} & m_{13} \\ m_{21} & m_{22} & m_{23} \\ m_{31} & m_{32} & m_{33} \end{bmatrix}$$

If the matrix $\mathbf{M}$ is positive definite, then the above matrix multiplication results in six linearly independent equations.

1.5.4 Summary of Matrix Methods

1. Bisection method for the roots of a polynomial.
2. MATLAB command *roots* for finding the roots of a polynomial.
3. Cholesky decomposition.

1.6 Approximation Methods for the Fundamental Frequency

The smallest natural frequency, known as the fundamental frequency, of a multi-DoF system is often of greater interest than the high natural frequencies because its forced response in many cases is the largest. One approach to this problem is to extend the Rayleigh method to matrix problems. We will see that the Rayleigh frequency approaches the fundamental frequency from the high side.

1.6.1 Rayleigh Method

Let $\mathbf{M}$ and $\mathbf{K}$ be the mass and stiffness matrices, respectively, and $\mathbf{x}$ is the assumed displacement vector for the amplitude of vibration. For harmonic motion, the maximum kinetic energy is

$$\mathbf{T}_{max} = 1/2\omega\mathbf{x}^t\mathbf{M}\mathbf{x}$$

and the maximum potential energy is

$$\mathbf{U}_{max} = 1/2\mathbf{x}^t\mathbf{K}\mathbf{x}$$

Since the maximum kinetic energy equals the maximum potential energy, these two quantities are equal. Equating these two and solving for ω^2 gives the Rayleigh quotient:

$$\omega^2 = \frac{\mathbf{x}^t\mathbf{K}\mathbf{x}}{\mathbf{x}^t\mathbf{M}\mathbf{x}}$$

It can be shown (Thomson and Dahleh, 1998) that this quotient approaches the lowest natural frequency from above and it is somewhat insensitive to the choice of amplitudes.

1.6.2   Dunkerley's Formula

Dunkerley's formula produces a lower bound for the fundamental frequency and can be used in conjunction with the Rayleigh method to get a good approximation for the fundamental frequency. Dunkerley's formula is based on the characteristic equation for the flexibility coefficients. The flexibility influence coefficient, a_{ii}, is defined as the displacement at i due to a unit force applied at j with all other forces equal to zero. This concept is most easily understood through an example.

1.6.2.1   Computation of the Flexibility Matrix

The procedure for computing the flexibility matrix and in particular, the computation of the flexibility matrix for the three spring–mass matrix systems shown in Figure 1.6, are discussed.

Example

First, one applies a unit force to mass 1 with no other forces present, i.e., $f_1 = 1$, $f_2 = f_3 = 0$. The displacements are located in the first column of the flexibility matrix.
This gives

$$\begin{bmatrix} x_1 \\ x_2 \\ x_3 \end{bmatrix} = \begin{bmatrix} 1/k_1 & 0 & 0 \\ 1/k_1 & 0 & 0 \\ 1/k_1 & 0 & 0 \end{bmatrix} \begin{bmatrix} 1 \\ 0 \\ 0 \end{bmatrix}$$

In this case, springs k_2 and k_3 are not stretched and are displaced equally with mass 1. Now, a unit force is applied to mass 2 and there are no other forces. This allows us to write the second column of the matrix to get

$$\begin{bmatrix} x_1 \\ x_2 \\ x_3 \end{bmatrix} = \begin{bmatrix} 0 & 1/k_1 & 0 \\ 0 & 1/k_1 + 1/k_2 & 0 \\ 0 & 1/k_1 + 1/k_2 & 0 \end{bmatrix} \begin{bmatrix} 0 \\ 1 \\ 0 \end{bmatrix}$$

This time, the unit force is transmitted through k_1 and k_2. The spring k_3 is not stretched. Finally, the force is applied to mass 3 and there are no other forces present. This gives the third column of the matrix:

$$\begin{bmatrix} x_1 \\ x_2 \\ x_3 \end{bmatrix} = \begin{bmatrix} 0 & 0 & 1/k_1 \\ 0 & 0 & 1/k_1 + 1/k_2 \\ 0 & 0 & 1/k_1 + 1/k_2 + 1/k_3 \end{bmatrix} \begin{bmatrix} 1 \\ 0 \\ 0 \end{bmatrix}$$

Since the flexibility matrix is the sum of the three previous matrices, it is given by

$$\begin{bmatrix} x_1 \\ x_2 \\ x_3 \end{bmatrix} = \begin{bmatrix} 1/k_1 & 1/k_1 & 1/k_1 \\ 1/k_1 & 1/k_1 + 1/k_2 & 1/k_1 + 1/k_2 \\ 1/k_1 & 1/k_1 + 1/k_2 & 1/k_1 + 1/k_2 + 1/k_3 \end{bmatrix} \begin{bmatrix} f_1 \\ f_2 \\ f_3 \end{bmatrix}$$

An interesting feature of the flexibility matrix is that it is symmetric about the diagonal. For simplicity of notation, let the ij element of the flexibility matrix be given by $a_{ij}m_j$. Dunkerley's formula is obtained from the characteristic equation of the flexibility matrix, which is obtained by computing the determinant of the following matrix:

$$\begin{vmatrix} a_{11}m_1 - 1/\omega^2 & a_{12} & a_{13} \\ a_{21} & a_{22}m_2 - 1/\omega^2 & a_{23} \\ a_{31} & a_{32} & a_{33}m_3 - 1/\omega^2 \end{vmatrix} = 0$$

A third-degree equation in $1/\omega^2$ is obtained by expanding the determinant. One obtains the following cubic equation:

$$(1/\omega^2)^3 - (a_{11}m_1 + a_{22}m_2 + a_{33}m_3)(1/\omega^2)^2 + \cdots = 0$$

A cubic equation has three roots that are denoted by $(1/\omega_i^2)$ for $i = 1, 2, 3$. This allows the cubic equation to be factored:

$$(1/\omega^2 - 1/\omega_1^2)(1/\omega^2 - 1/\omega_2^2)(1/\omega^2 - 1/\omega_3^2) = 0$$

The highest two powers of this equation are given by

$$(1/\omega^2)^3 - (1/\omega_1^2 + 1/\omega_2^2 + 1/\omega_3^2)(1/\omega^2)^2 + \cdots = 0$$

The coefficient of the second highest power is equal to the sum of the roots of the characteristic equation, which is also equal to the sum of the diagonal elements of the matrix $\mathbf{A}^{-1}$. This relationship is not just true for $n = 3$ but is more generally true for n greater than or equal to 3. For the general n-DoF system

$$1/\omega_1^2 + 1/\omega_2^2 + \cdots 1/\omega_n^2 = a_{11}m_1 + a_{22}m_2 + \cdots + a_{nn}m_n$$

The fundamental frequency is the smallest natural frequency. Since $\omega_2, \omega_3, \ldots$ are larger than ω_1, the reciprocal of these frequencies is smaller then the reciprocal of the fundamental frequency. An estimate for the fundamental frequency is obtained by neglecting all of the higher modes in the left-hand side of the above equation. This estimate gives a value for ω_1 that is smaller then the true value of the fundamental frequency. Dunkerley's formula is a lower bound for the fundamental frequency and it is given by

$$1/\omega_1^2 < a_{11}m_1 + a_{22}m_2 + \cdots + a_{nn}m_n$$

1.6.3   Summary of Approximations for the Fundamental Frequency

1. Rayleigh method

$$\omega^2 = \frac{\mathbf{x}^t\mathbf{K}\mathbf{x}}{\mathbf{x}^t\mathbf{M}\mathbf{x}}$$

2. Dunkerley's formula $1/\omega_1^2 < a_{11}m_1 + a_{22}m_2 + \cdots + a_{nn}m_n$.

1.7   Finite Element Method

In the finite element method, complex structures are replaced by assemblages of simple structural elements known as finite elements. The elements are connected by joints or nodes. The force and moments at the ends of the elements are known from structural theory, the joints between the elements are matched for compatibility of displacement, and the force and moment at the joints are established by imposing the condition of equilibrium.

The accuracy obtainable from the finite element method depends on being able to duplicate the vibration mode shapes. Using only one finite element between structure joints or corners gives good results for the first lowest mode, because the static deflection curve is a good approximation to the lowest dynamic mode shape. For higher modes, several elements are necessary between structural joints. This leads to large matrices. The eigenvalues and eigenvectors need to be computed numerically.

This section introduces the basic idea of the finite element method as it applies to the simple vibration problem.

The basic idea behind the finite element method is to break up the structure into simple component structures. The structural elements for the bar and the beam are discussed here.

1.7.1 Bar Element

The force–displacement relationship for a uniform rod is

$$F = (EA/L)U$$

where E is the young's modulus, A is the cross sectional area, L is the length of the element and U is the displacement. Figure 1.8 shows this one-dimensional element. The two endpoints are the nodes.

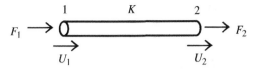
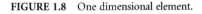

FIGURE 1.8 One dimensional element.

For simplicity, assume the axial displacement at any point $\varsigma = x/L$ is linear:

$$U(x, t) = a(t) + b(t)x$$

and $U_1(t) = U(0, t)$, $U_2(t) = U(L, t)$. These two conditions uniquely determine the coefficients $a(t)$ and $b(t)$. They are given by the following:

$$a(t) = U_1(t)$$

and

$$b(t) = (U_2(t) - U_1(t))/L$$

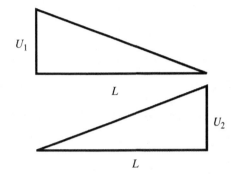

FIGURE 1.9 Linear mode shapes.

Using the linear element, the displacement anywhere along the beam is given by

$$U(x, t) = (1 - x/L)U_1(t) + x/LU_2(t)$$

$$= \varphi_1 U_1(t) + \varphi_2 U_2(t)$$

where

$$\varphi_1 = 1 - x/L \quad \text{and} \quad \varphi_2 = x/L$$

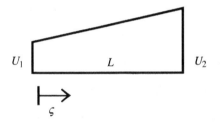
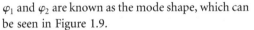

FIGURE 1.10 Superposition of the linear mode shapes.

φ_1 and φ_2 are known as the mode shape, which can be seen in Figure 1.9.

These two mode shapes can be superimposed to create a linear function. An example of such a function is given in Figure 1.10.

The kinetic energy of the bar is given by

$$T = .5\int_0^l \dot{u}^2 m dx = .5m \int_0^l ((1 - \varsigma)\dot{u}_1(t) + \varsigma\dot{u}_2(t))^2 l d\varsigma = .5ml(1/3\dot{u}_1^2 + 1/3\dot{u}_1\dot{u}_2 + 1/3\dot{u}_2^2)$$

1.7.1.1 Mass Matrix

The generalized mass matrix is derived from the Lagrange equations using the following:

$$\frac{D}{Dt}\frac{\partial T}{\partial \dot{u}_1}$$

Given the kinetic energy for the bar, the Lagrange equations become

$$\frac{D}{Dt}\frac{\partial T}{\partial \dot{u}_1} = mL(1/3\ddot{u}_1 + 1/6\ddot{u}_2)$$

$$\frac{D}{Dt}\frac{\partial T}{\partial \dot{u}_2} = mL(1/3\ddot{u}_1 + 1/6\ddot{u}_2)$$

The mass matrix for an axial element with a uniform mass distribution per unit length is given by

$$\frac{mL}{6}\begin{bmatrix} 2 & 1 \\ 1 & 2 \end{bmatrix}$$

1.7.1.2   Stiffness Matrix

The force–displacement relationship for a uniform bar is given by

$$\begin{bmatrix} F_1 \\ F_2 \end{bmatrix} = \frac{EA}{L}\begin{bmatrix} 1 & -1 \\ -1 & 1 \end{bmatrix}\begin{bmatrix} u_1 \\ u_1 \end{bmatrix}$$

1.7.1.3   Variable Properties

A simple approach to problems with variable properties is to use a large number of elements of short length. The variation of mass or stiffness over each element is small and can be neglected. In this model, the mass and stiffness for each element is constant and can be placed outside the integral. If a large number of elements are needed to capture the behavior, this will lead to a large matrix problem.

Example

A tapered rod is modeled as two uniform sections where $EA_1 = 2EA_2$ and $m_1 = 2m_2$, $k_2 = 2EA_2/L$. The displacement equation for this system is given by

$$\frac{-\omega^2 m_2}{6}\begin{bmatrix} 2(2+1) & 1 \\ 1 & 2 \end{bmatrix} + \frac{2EA_2}{L}\begin{bmatrix} 2+1 & -1 \\ -1 & 1 \end{bmatrix}\begin{bmatrix} u_1 \\ u_2 \end{bmatrix} = \begin{bmatrix} 0 \\ 0 \end{bmatrix}$$

This can be written as a standard eigenvalue problem where $\lambda = \omega^2 m_2 L/12 EA_2$.
    The eigenvalue problem is the following

$$\begin{vmatrix} 3 - 6\lambda & -(1+\lambda) \\ -(1+\lambda) & (1-2\lambda) \end{vmatrix} = 0$$

The solution of the determinant requires computing the roots of a quadratic equation. The two roots are $\lambda = [0.6140, 1.1088]$. The two natural frequencies can be computed from λ. They are

$$\omega_1 = 1.4029\sqrt{((EA_2)/(M_2 L))}$$

$$\omega_2 = 3.6477\sqrt{((EA_2)/(M_2 L))}$$

The modes shapes are calculated by solving the eigenvalue problem. The two eigenvectors are

$$\mathbf{x}_1 = [0.5773, 1]^t$$

and

$$\mathbf{x}_2 = [-5258, 1]^t$$

1.7.2   Beam

When the ends of the element are rigidly connected to the adjoining structure, the elements act like a beam with the moments and lateral forces acting at the joints. Generally, the axial displacement $u_2 - u_1$ is small compared to the lateral displacement V of the beam.
    The local coordinates of the beam element are lateral displacements, V, and rotation, θ, at the two ends. This results in four coordinates V_1, θ_1, V_2, θ_2. Four constraints uniquely determine a cubic polynomial. Therefore, the lateral displacement of a beam is assumed to be described by a cubic

polynomial

$$V(x) = p_1 + p_2 s + p_3 s^2 + p_4 s^4$$

where $s = x/L$ and p_i, the coefficients of the polynomials, are related to the lateral displacement and the rotation. The lateral displacement at the left-hand side determines V_1:

$$V(0, t) = V_1(t) = p_1$$

The rotation at the left-hand end determines the second coefficient, θ_1:

$$\frac{\partial V}{\partial x}(0, t) = \theta_1 = p_2$$

The remaining two coefficients are combinations of the variables and they are given by the following two constraints:

$$V(L, t) = V_2(t) \text{ and } \frac{\partial V}{\partial x}(L, t) = \theta_2$$

Applying these two conditions, one obtains

$$P_3 = 1/L^2(-3v_1(t) - 2\theta_1(t)L + 3v_2(t) - \theta_2(t)L)$$

$$P_4 = 1/L^3(2v_1(t) + \theta_1(t)L - 2v_2(t) + \theta_2(t)L)$$

These coefficients can be represented by the following matrix equation:

$$\begin{bmatrix} p_1 \\ p_2 \\ p_3 \\ p_4 \end{bmatrix} = \begin{bmatrix} 1 & 0 & 0 & 0 \\ 0 & 1 & 0 & 0 \\ -3 & -2 & 3 & 1 \\ 2 & 1 & -2 & 1 \end{bmatrix} \begin{bmatrix} v_1 \\ L\theta_2 \\ v_2 \\ L\theta \end{bmatrix}$$

The shape functions for the beam elements are determined by equating a single displacement to one, and all other displacement to zero. The first shape function is derived by letting $V_1 = 1$ and the remaining three variables zero; i.e., $V_2 = \theta_1 = \theta_2 = 0$. This gives

$$p_1 = 1 \qquad p_2 = 0 \qquad p_3 = -3 \qquad p_4 = 2$$

The first shape function becomes

$$\varphi_1 = 1 - 3s^2 + 2s^3$$

A similar calculation for $\theta_1 = 1$ and $V_2 = V_1 = \theta_2 = 0$ yields

$$p_1 = 0 \qquad p_2 = l \qquad p_3 = -2l \qquad p_4 = l$$

and the shape function becomes

$$\varphi_2 = ls - 2ls^2 + ls^3$$

The remaining two shape functions are determined similarly. The four shape functions for the beam are

$$\varphi_1 = 1 - 3s^2 + 2s^3$$

$$\varphi_2 = ls - 2ls^2 + ls^3$$

$$\varphi_3 = 3s^2 - 2s^3$$

$$\varphi_4 = -ls^2 + ls^3$$

1.7.2.1   Mass Matrix

Just as in the case of the bar, the generalized mass m_{ij} is given by

$$M_{ij} = \int_0^l \varphi_i \varphi_j m \mathrm{d}x$$

Substitution of the above shape functions and integration yields the following mass matrix for the uniform beam element in terms of the end displacements:

$$\frac{mL}{420} \begin{bmatrix} 156 & 22L & 54 & -13L \\ 22L & 2L^2 & 13L & -3L^2 \\ 54 & 13L & 156 & -22L \\ -13L & -3L^2 & -22L & 4L^2 \end{bmatrix}$$

1.7.3   Summary of Finite Element Method

1. Linear finite element for a bar $\varphi_1 = 1 - x/L$ and $\varphi_2 = x/L$
2. Mass matrix for the bar

$$\frac{ML}{6} \begin{bmatrix} 2 & 1 \\ 1 & 2 \end{bmatrix}$$

3. Stiffness matrix for the bar

$$\begin{bmatrix} F_1 \\ F_2 \end{bmatrix} = \frac{EA}{L} \begin{bmatrix} 1 & -1 \\ -1 & 1 \end{bmatrix} \begin{bmatrix} u_1 \\ u_2 \end{bmatrix}$$

4. Cubic finite elements for the beam

$$\varphi_1 = 1 - 3s^2 + 2s^3$$

$$\varphi_2 = ls - 2ls^2 + ls^3$$

$$\varphi_3 = 3s^2 - 2s^3$$

$$\varphi_4 = -ls^2 + ls$$

Appendix 1A
Introduction to MATLAB®

MATLAB is a software package for numerical computation, visualization, and symbolic manipulation. It is an interactive environment with hundreds of built-in functions, which are in essence subroutines. These functions range from plotting commands, to those for finding the eigenvalues and eigenvectors of a matrix, to those for solving an ordinary differential equation, and much more. In addition to the built-in functions, MATLAB contains a programming language, which allows the user to write their own functions. The name MATLAB is an abbreviation of *matrix laboratory*. The original versions of MATLAB concentrated on numerical analysis of linear systems of equation. MATLAB is available from the Mathworks (www.mathworks.com).

The best way to learn MATLAB is by playing around with the different functions (Pratap, 2001). The first thing to note when you launch MATLAB is that it is a window-based environment. There are three windows: the command window, the graphics window, and the edit window. The command window is the main window and it is the one in which you run all functions (built-in or user created). This is the window which appears when the program is launched, and it has the symbol $\gg$ as a prompt. The graphics window is where all of the graphics are displayed, and the edit window is where users create and save all of their own programs, known as *m files*.

MATLAB provides routines for all of the basic areas of numerical analysis (numerical linear algebra, data analysis, Fourier transform, and interpolation), curve fitting, root-finding, numerical solution of ordinary differential equations, integration, and graphics. There are specialized tool boxes for signal processes and control systems to name two. With these hundreds of functions, it is imperative to have a good help facility. In the command window, the command *help functionname* provides online help. For example, type *help help* to get more information about help. There are three other commands for information: *lookfor*, *helpwin*, and *helpdesk*. *Lookfor* gives a list of functions with the keyword in their description. *Helpwin* gives a help window. *Helpdesk* is a web browser-based help.

Given the early history of MATLAB as a matrix laboratory, it should not be surprising that one of its strengths is its ability to manipulate vectors and matrices very well. A row vector is created by typing in the command window the following:

$$\gg x = [2, 3, 6, 4]$$

This command will produce

$$x = 2 \quad 3 \quad 6 \quad 4$$

A common vector is created by entering the following

$$\gg x = [2; 1; 3; 4]$$

This produces $x =$

$$2$$

$$1$$

$$3$$

$$4$$

The elements of a vector or matrix are separated by commas or by spaces, and the rows are separated by semicolons. Printing is suppressed by ending the line with a semicolon. The following command will produce a 2 × 2 matrix but it will not print it:

$$A = \begin{bmatrix} 2 & 4 \\ 1 & 5 \end{bmatrix}$$

Providing that the operations make sense, it is easy to do operations on vectors and matrices in MATLAB. For example, if **A** and **B** are two matrices of the same size, the command $A + B$ adds the matrices.

There are several MATLAB commands given throughout the text of this chapter. Several books are available to get you started with MATLAB; for example, Hanselman and Littlefield (2001), Palm (2001), Pratap (2001), and Recktenwald (2000). The best way to learn more about these commands is to type help and the command name. This way, you get the most up-to-date information concerning the function.

References

Atkinson, K. 1978. *An Introduction to Numerical Analysis*, 2nd ed., Wiley, New York.

Cheney, E. and Kincaid, D. 1999. *Numerical Mathematics and Computing*, 4th ed., Brooks Cole, Monterey, CA.

Hanselman, D. and Littlefield, B. 2001. *Mastering MATLAB 6: A Comprehensive Tutorial and Reference*, Prentice Hall, Upper Saddle River, NJ.

Isaacson, E. and Keller, H. 1966. *Analysis of Numerical Methods*, Wiley, New York.

Palm, W. 2001. *Introduction to MATLAB 6 for Engineers*, McGraw Hill, Boston.

Pratap, R. 2001. *Getting Started with MATLAB 6: A Quick Introduction for Scientists and Engineers*, Oxford University Press, New York.

Recktenwald, G. 2000. *Numerical Methods with MATLAB: Implementation and Application*, Prentice Hall, Upper Saddle River, NJ.

Strang, G. 1986. *Introduction to Applied Mathematics*, Wellesley-Cambridge Press, Wellesley, MA.

Thomson, W. and Dahleh, M. 1998. *Theory of Vibration with Applications*, 5th ed., Prentice Hall, Upper Saddle River, NJ.

2

Vibration Modeling and Software Tools

Datong Song
National Research Council of Canada

Cheng Huang
National Research Council of Canada

Zhong-Sheng Liu
National Research Council of Canada

Summary

In this chapter, several aspects of vibration modeling are addressed. They include the formulation of the equations of motion both in differential form and integral form, the Rayleigh–Ritz method and the finite element methods, and model reduction. Natural vibration analysis and response analysis are discussed in detail. Several commercial finite element analysis (FEA) software tools are listed and their capabilities for vibration analysis are introduced. The basic procedure in using the commercial FEA software packages for vibration analysis is outlined (also see Chapter 1 and Chapter 4). The vibration analysis of a gearbox housing is presented to illustrate the procedure.

2.1 Introduction

Vibration phenomenon, common in mechanical devices and structures [2,9], is undesirable in many cases, such as machine tools. But this phenomenon is not always unwanted; for example, vibration is needed in the operation of vibration screens. Thus, reducing or utilizing vibration is among the challenging tasks that mechanical or structural engineers have to face. Vibration modeling has been used extensively for a better understanding of vibration phenomena. The vibration modeling here implies a process of converting an engineering vibration problem into a mathematical model, whereby the major vibration characteristics of the original problem can be accurately predicted. The mathematical model of vibration in its general sense consists of four components: a mass (inertia) term; a stiffness term; an excitation force term; and a boundary condition term. These four terms are represented in differential equations of motion for discrete (or, lumped-parameter) systems, or boundary value problems for continuous systems. A damping term is included if damping effects are of concern. Depending on the nature of the vibration problem, the complexity of the mathematical model varies from simple spring–mass systems to multi-degree-of-freedom (DoF) systems; from a continuous system for a single structural member (beam, rod, plate, or shell) to a combined system for a built-up structure; from a linear system to a nonlinear system. The success of the mathematical model heavily depends on whether or not the four terms mentioned before can represent the actual vibration problem. In addition, the mathematical model must be sufficiently simplified in order to produce an acceptable computational cost. The construction of such a representative and simple mathematical model requires an in-depth understanding of vibration principles and techniques, extensive experience in vibration modeling, and ingenuity in using vibration software tools. Furthermore, it also requires sufficient knowledge of the vibration problem itself in terms of working conditions and specifications.

Except for few special cases that promise exact and explicit analytical solutions, vibration models have to be studied by means of approximate numerical methods such as the finite element method. The finite element method has been very successfully used for vibration modeling for the past two decades. Its success is attributed to the development of sophisticated software packages and the rapid growth of computer technology.

In this chapter, several aspects of the construction of mathematical models of linear vibration problems without damping will be addressed. The capabilities of the available software packages for vibration analysis are listed and the basic procedure for vibration analysis is summarized. As an illustration, an engineering example is given.

2.2   Formulation

2.2.1   Differential Formulation

In a majority of engineering vibration problems, the amplitude of vibrations is very small, so that the following assumptions hold: (1) a linear form of strain–displacement relationships, and (2) a linear form of stress–strain relationships (Hooke's Law). If the energy losses are negligible, it is straightforward to apply Newton's (second) law and Hooke's Law to derive the equations of motion, which appear as differential equations. Consider a single-DoF spring–mass system, as shown in Figure 2.1. The two laws are given by

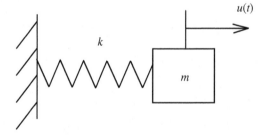

$$\begin{cases} m\ddot{u}(t) = -f, & \text{Newton's law} \\ f = ku(t), & \text{Hooke's Law} \end{cases} \quad (2.1)$$

FIGURE 2.1   Single-DoF spring–mass system.

The first equation describes the inertia force, and the second equation describes the elastic force. These two forces are essential for mechanical vibration to exist.

In a similar way, the differential equations are given directly when Newton's law plus Hooke's Law is applied to a multiple-DoF spring–mass system, shown in Figure 2.2

$$\begin{cases} \mathbf{M}\ddot{u}(t) = -\mathbf{F}, & \text{Newton's law} \\ \mathbf{F} = \mathbf{K}u(t), & \text{Hooke's Law} \end{cases} \quad (2.2)$$

where $\mathbf{M}$ is the (diagonal) mass matrix, and $\mathbf{K}$ is the stiffness matrix.

In the case of continua, the differential equations of motion can be derived by means of Newton's law and Hooke's Law in the same way as above. But in this case the boundary conditions have to be specified

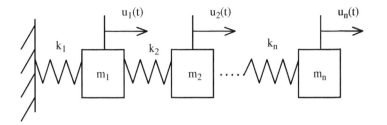

FIGURE 2.2 Multiple-DoF spring–mass system.

in order to make the problem statement complete. For example, as a direct consequence of Newton's law and Hooke's Law, the differential equation of bending vibration of a clamped–clamped Euler beam may be given as

$$
\begin{cases}
\rho \dfrac{\partial^2 u(x,t)}{dt^2} = -f, & \text{Newton's law} \\[2ex]
f = EI \dfrac{\partial^4 u(x,t)}{\partial x^4}, & \text{Hooke's Law} \\[2ex]
u(0,t) = u(l,t) = \dfrac{\partial u(0,t)}{\partial x} = \dfrac{\partial u(l,t)}{\partial x} = 0, & \text{Boundary conditions}
\end{cases}
\tag{2.3}
$$

where ρ represents the mass per unit length, l the beam length, and EI the bending stiffness (flexural rigidity).

2.2.2 Integral Formulation and Rayleigh–Ritz Discretization

Besides the approach in which Newton's law and Hooke's Law are directly used to establish equations of motion, there are alternatives: Hamilton's principle, the minimum potential energy principle, and the virtual work principle; all of which appear in integral form. From a mathematical standpoint, the differential equations and the integral equations are equivalent in that one can be derived from another. However, they are very different in that the integral equations facilitate the application of the discretization schemes such as the finite element method, an element-wise application of Rayleigh–Ritz method. Therefore, Hamilton's principle, as one of the integral formulations, and its Rayleigh–Ritz discretization are briefly introduced here in order to provide a better understanding of the finite element method.

Denote T as the system kinetic energy, V the system potential energy, and δW the virtual work done by nonconservative forces. Hamilton's principle [11] states that the variation of the Lagrangian $(T - V)$ Standard terminology plus the line integral of the virtual work done by the nonconservative forces during any time interval must be equal to zero. If the time interval is denoted by $[t_1, t_2]$, then Hamilton's principle can be expressed as

$$
\delta \int_{t_1}^{t_2} (T - V)dt + \int_{t_1}^{t_2} \delta W \, dt = 0
\tag{2.4}
$$

In the case of a continuum, we look for an approximate solution $u(x, y, z, t)$ in the form of

$$
u(x, y, z, t) = \sum_{i=1}^{n} \varphi_i(x, y, z) q_i(t)
\tag{2.5}
$$

where $\varphi_i(x, y, z)$ is called a Rayleigh–Ritz shape function and $q_i(t)$ is called a generalized coordinate. In this way, the system kinetic energy and the system potential energy can be, respectively, expressed as follows:

$$T = \frac{1}{2} \sum_{i=1}^{n} \sum_{j=1}^{n} m_{ij} \dot{q}_i \dot{q}_j \equiv \frac{1}{2} \dot{\mathbf{u}}^{\mathrm{T}}(t) \mathbf{M} \dot{\mathbf{u}}(t) \tag{2.6}$$

and

$$V = \frac{1}{2} \sum_{i=1}^{n} \sum_{j=1}^{n} k_{ij} q_i q_j \equiv \frac{1}{2} \mathbf{u}^{\mathrm{T}}(t) \mathbf{K} \mathbf{u}(t) \tag{2.7}$$

where $\mathbf{u}^{\mathrm{T}}(t) \equiv [q_1, q_2, \ldots, q_n]$, $\mathbf{M} \equiv [m_{ij}]$, $\mathbf{K} \equiv [k_{ij}]$.

The virtual work done by the generalized forces is

$$\delta W = \sum_{i=1}^{n} f_i(t) \delta q_i = \mathbf{F}^{\mathrm{T}} \, \delta \mathbf{u}(t) \tag{2.8}$$

where $\mathbf{F}^{\mathrm{T}} \equiv [f_1(t), f_2(t), \ldots, f_n(t)]$ and $f_i(t)$ is the generalized force corresponding to the nonconservative force $f(x, y, z, t)$

$$f_i(t) = \int \varphi_i(x, y, z) f(x, y, z, t) \mathrm{d}v \tag{2.9}$$

Substituting Equation 2.6, Equation 2.7 and Equation 2.8 into Hamilton's principle (Equation 2.4) and conducting a routine variation operation, one has

$$\int_{t_1}^{t_2} (\dot{\mathbf{u}}^{\mathrm{T}} \mathbf{M} \, \delta \dot{\mathbf{u}} - \mathbf{u}^{\mathrm{T}} \mathbf{K} \, \delta \mathbf{u} + \mathbf{F}^{\mathrm{T}} \, \delta \mathbf{u}) \mathrm{d}t = 0 \tag{2.10}$$

Applying the separation integration to the first term of the above equation and noting that the variations of the generalized coordinate $\delta \mathbf{u}$ at times t_1 and t_2 equal zero, Equation 2.10 is rewritten as

$$\int_{t_1}^{t_2} (-\mathbf{M} \ddot{\mathbf{u}}(t) - \mathbf{K} \mathbf{u}(t) + \mathbf{F}) \delta \mathbf{u} \, \mathrm{d}t = 0 \tag{2.11}$$

Because $\delta \mathbf{u}$, the variation of the generalized coordinate vector, is arbitrary and independent, from the above equation one obtains

$$\mathbf{M} \ddot{\mathbf{u}}(t) + \mathbf{K} \mathbf{u} = \mathbf{F} \tag{2.12}$$

which is the vibration equation resulting from a Rayleigh–Ritz discretization.

2.2.3   Finite Element Method

In the finite element method (FEM) [7,10,12], a continuum is divided into a number of relatively small regions called elements that are interconnected at selected nodes (also see Chapter 4). This procedure is called discretization. The deformation within each element is expressed by interpolating polynomials. The coefficients of these polynomials are defined in terms of the element nodal DoF that describe the displacements and slopes of selected nodes on the element. By using the connectivity between elements, the assumed displacement field can then be written in terms of the nodal DoF by means of the element shape function. Using the assumed displacement field, the kinetic energy and the strain energy of each element are expressed in the form of the element mass and stiffness matrices. The energy expressions for the entire continua can be obtained by adding the energy expressions of its elements. This leads to the assembled mass matrix and the assembled stiffness matrix, and finally to the finite element vibration equation.

The displacement in the interior of an element e is determined by a polynomial

$$u(x, y, z, t) = \mathbf{N} \mathbf{u}^e \tag{2.13}$$

where the matrix $\mathbf{N}$ is called the shape function matrix of the element e, and $\mathbf{u}^e$ the vector of the nodal DoF.

Based on the element displacement expression Equation 2.13, one can obtain the strain and the stress in the element e and finally the strain energy.

The strain and the stress in the element e are

$$\boldsymbol{\varepsilon} = \partial u(x, y, z, t) = \partial \mathbf{N} \mathbf{u}^e = \mathbf{B} \mathbf{u}^e \tag{2.14}$$

and

$$\boldsymbol{\sigma} = \mathbf{D} \boldsymbol{\varepsilon} = \mathbf{D} \mathbf{B} \mathbf{u}^e = \mathbf{S} \mathbf{u}^e \tag{2.15}$$

respectively, where ∂ is the differential operator matrix, $\mathbf{B} = \partial \mathbf{N}$ is the element strain matrix, $\mathbf{D}$ is the elastic matrix, and $\mathbf{S} = \mathbf{DB}$ is called the element stress matrix.

The strain energy in the element e is given by the element strain $\boldsymbol{\varepsilon}$ and stress $\boldsymbol{\sigma}$ as

$$V^e = \frac{1}{2} \int \boldsymbol{\varepsilon}^{\mathrm{T}} \boldsymbol{\sigma} \, \mathrm{d}v = \frac{1}{2} (\mathbf{u}^e)^{\mathrm{T}} \mathbf{K}^e \mathbf{u}^e \tag{2.16}$$

where

$$\mathbf{K}^e = \int \mathbf{B}^{\mathrm{T}} \mathbf{D} \mathbf{B} \, \mathrm{d}v \tag{2.17}$$

is called the element stiffness matrix.

The velocity at a point (x, y, z) in the element e can be obtained from Equation 2.13 as

$$\dot{u}(x, y, z, t) = \mathbf{N} \dot{\mathbf{u}}^e \tag{2.18}$$

So the kinetic energy of the element e is

$$T^e = \frac{1}{2} \int \rho \dot{\mathbf{u}}^{\mathrm{T}} \dot{\mathbf{u}} \, \mathrm{d}v = \frac{1}{2} (\dot{\mathbf{u}}^e)^{\mathrm{T}} \mathbf{M}^e \dot{\mathbf{u}}^e \tag{2.19}$$

where

$$\mathbf{M}^e = \int \rho \mathbf{N}^{\mathrm{T}} \mathbf{N} \, \mathrm{d}v \tag{2.20}$$

is called the element mass matrix.

The equivalent nodal force $\mathbf{F}^e$ corresponding to the force $\mathbf{f}^e$ applied onto the element e is determined by equaling the work done by $\mathbf{F}^e$ to the work done force by $\mathbf{f}^e$ along any virtual displacement. This leads to the following:

$$(\delta \mathbf{u}^e)^{\mathrm{T}} \mathbf{F}^e = \int \delta \mathbf{u}^{\mathrm{T}} \mathbf{f}^e \, \mathrm{d}v = (\delta \mathbf{u}^e)^{\mathrm{T}} \left(\int \mathbf{N}^{\mathrm{T}} \mathbf{f}^e \, \mathrm{d}v \right) \tag{2.21}$$

Note that as the variation of the nodal displacement is arbitrary, one can obtain the expression of the equivalent nodal force $\mathbf{F}^e$ from Equation 2.21 as

$$\mathbf{F}^e = \int \mathbf{N}^{\mathrm{T}} \mathbf{f}^e \, \mathrm{d}v \tag{2.22}$$

Now we have the kinetic energy, the strain energy, and the equivalent nodal force of the element e. But these quantities are expressed in the local coordinate system (X^e, Y^e, Z^e) of the element e, not in the global coordinate system (X, Y, Z). In order to calculate the corresponding counterparts for the whole structure, it is necessary to transform the expressions of the kinetic energy, the strain energy, and the equivalent nodal force of the element e from the local coordinate system into the global one.

Let $\mathbf{L}$ be the transformation matrix from the global coordinate system to the local coordinate system. Then the nodal displacement vector $\mathbf{u}^e$ in the local coordinate system is related to the nodal displacement vector $\bar{\mathbf{u}}^e$ in the global coordinate system by the following:

$$\mathbf{u}^e = \mathbf{L}\bar{\mathbf{u}}^e \tag{2.23}$$

Similarly, the equivalent nodal force vector $\mathbf{F}^e$ in the local coordinate system is related to the equivalent nodal force vector $\bar{\mathbf{F}}^e$ in the global coordinate system by

$$\mathbf{F}^e = \mathbf{L}\bar{\mathbf{F}}^e \tag{2.24}$$

Substituting Equation 2.23 into Equation 2.16 and Equation 2.19, and noting that $\mathbf{L}$ is a normal matrix ($\mathbf{L}^T = \mathbf{L}^{-1}$), the element stiffness and mass matrices in the global coordinate system can be, respectively, expressed as

$$\bar{\mathbf{K}}^e = \mathbf{L}^T\mathbf{K}^e\mathbf{L} \tag{2.25}$$

and

$$\bar{\mathbf{M}}^e = \mathbf{L}^T\mathbf{M}^e\mathbf{L} \tag{2.26}$$

The equivalent nodal force vector in the global coordinate system is solved from Equation 2.24

$$\bar{\mathbf{F}}^e = \mathbf{L}^T\mathbf{F}^e \tag{2.27}$$

In this way, we can obtain the total strain energy of the structure as

$$V = \sum_e \bar{V}^e = \frac{1}{2}\sum_e (\bar{\mathbf{u}}^e)^T\bar{\mathbf{K}}^e\bar{\mathbf{u}}^e = \frac{1}{2}\mathbf{u}^T\mathbf{K}\mathbf{u} \tag{2.28}$$

where the matrix

$$\mathbf{K} = \sum_e \bar{\mathbf{K}}^e \tag{2.29}$$

is called the global stiffness matrix of the structure. The vector $\mathbf{u}$ is the global nodal displacement vector of the structure.

Similarly, the total kinetic energy of all of the elements can be written as

$$T = \sum_e \bar{T}^e = \frac{1}{2}\sum_e (\dot{\bar{\mathbf{u}}}^e)^T\bar{\mathbf{M}}^e\dot{\bar{\mathbf{u}}}^e = \frac{1}{2}\dot{\mathbf{u}}^T\mathbf{M}\dot{\mathbf{u}} \tag{2.30}$$

where the matrix

$$\mathbf{M} = \sum_e \bar{\mathbf{M}}^e \tag{2.31}$$

is called the global mass matrix. The vector $\dot{\mathbf{u}}$ is the global nodal velocity vector.

The total virtual work done by the external forces is

$$\delta W = \sum_e \delta\bar{W}^e = \sum_e (\delta\bar{\mathbf{u}}^e)^T\bar{\mathbf{F}}^e = (\delta\mathbf{u})^T\mathbf{F} \tag{2.32}$$

where the vector

$$\mathbf{F} = \sum_e \bar{\mathbf{F}}^e \tag{2.33}$$

is a global generalized force vector.

Substituting Equation 2.28, Equation 2.30, and Equation 2.32 into Hamilton's principle (Equation 2.4) and conducting the routine variation operation, one has

$$\mathbf{M\ddot{u} + Ku = F} \tag{2.34}$$

which is the vibration equation resulting from the finite element discretization.

2.2.4 Lumped Mass Matrix

The element mass matrix given by Equation 2.20 is normally a full symmetric matrix, because the element shape functions are not orthogonal with each other. It is desirable to reduce this full matrix into a diagonal matrix. In practice, this is achieved by lumping the element mass at its nodes. For example, the consistent element mass matrix of a beam element is

$$\mathbf{M}^e = \frac{\rho A l}{420} \begin{bmatrix} 156 & 22l & 54 & -13l \\ 22l & 4l^2 & 13l & -3l^2 \\ 54 & 13l & 156 & -22l \\ -13l & -3l^2 & -22l & 4l^2 \end{bmatrix} \tag{2.35}$$

When the inertia effect associated with the rotational DoF is negligible, the element lumped mass matrix can be obtained by lumping one half of the total beam element mass at each of the two nodes along the translation DoF:

$$\mathbf{M}^e = \frac{\rho A l}{2} \begin{bmatrix} 1 & 0 & 0 & 0 \\ 0 & 0 & 0 & 0 \\ 0 & 0 & 1 & 0 \\ 0 & 0 & 0 & 0 \end{bmatrix} \tag{2.36}$$

When the inertia effect associated with the rotational DoF is not negligible, the mass moment of inertia of one half of the beam element about each node can be computed and included at the diagonal locations corresponding to the rotational DoF:

$$\mathbf{M}^e = \frac{\rho A l}{2} \begin{bmatrix} 1 & 0 & 0 & 0 \\ 0 & l^2/12 & 0 & 0 \\ 0 & 0 & 1 & 0 \\ 0 & 0 & 0 & l^2/12 \end{bmatrix} \tag{2.37}$$

2.2.5 Model Reduction

The finite element discretization of an engineering vibration problem usually generates a very large number of DoF. In particular, when automatic meshing schemes are not properly applied, or three-dimensional elements must be used, the number of elements created could become too great to be cost-effectively handled with limited computer capabilities. To solve this problem, modelers have to pay close attention to how the meshing is done in commercial software packages. Very often, simplification and idealization based on the nature of the problem of concern can tremendously reduce the number of elements. For example, there could be two ways of generating the finite elements of a clamped-free steel beam with a metal block attached to its free end. One way is to mesh both the beam and the block using three-dimensional elements; the other way is to mesh the beam with one-dimensional beam elements and treat the block as a lumped mass, zero-dimensional element. It is obvious that the first approach will result in many more elements than the second approach. However, both approaches will give very similar results for the first several natural frequencies and the associated mode shapes. Another technique for reducing the number of elements comes from deleting the detailed features. The detailed features here imply those geometrical details, such as filets, chamfers, small holes, and so on, which do not have significant contributions to the vibration behavior of the entire structure, but increase the number of elements. Generally these detailed features can be deleted without any visible effect on the results, if the global behavior of the vibration problem is of concern. Note that such detailed features may have to be kept if the localized behavior such as fatigue (stress) induced by vibration is to be evaluated.

When further model reduction is necessary, Guyan reduction [3] is considered. It was proposed two decades ago when computer capabilities were much more limited than today. In fact, Guyan reduction is still in use today and has been cast into many commercial software packages. In Guyan reduction, the model scale is reduced by removing those DoF (called slave DoF) that can be approximately expressed by the rest of the DoF (called master DoF) through a static relation. The DoF associated with zero mass or relatively small mass are likely candidates for slave DoF.

By rearranging the DoF $\mathbf{u}$ so that those to be removed, denoted by $\mathbf{u}^2$, appear last in the vector, and partitioning the mass and the stiffness matrices accordingly, one obtains

$$\begin{bmatrix} \mathbf{M}_{11} & \mathbf{M}_{12} \\ \mathbf{M}_{21} & \mathbf{M}_{22} \end{bmatrix} \begin{Bmatrix} \ddot{\mathbf{u}}^1 \\ \ddot{\mathbf{u}}^2 \end{Bmatrix} + \begin{bmatrix} \mathbf{K}_{11} & \mathbf{K}_{12} \\ \mathbf{K}_{21} & \mathbf{K}_{22} \end{bmatrix} \begin{Bmatrix} \mathbf{u}^1 \\ \mathbf{u}^2 \end{Bmatrix} = \begin{Bmatrix} \mathbf{F} \\ \mathbf{0} \end{Bmatrix} \qquad (2.38)$$

If we assume $\mathbf{M}_{22} = \mathbf{0}$, and $\mathbf{M}_{21} = \mathbf{0}$, then the second equation in Equation 2.38 can be written as

$$\mathbf{u}^2 = -\mathbf{K}_{22}^{-1}\mathbf{K}_{21}\mathbf{u}^1 \qquad (2.39)$$

Define the transformation

$$\mathbf{u} = \mathbf{Q}\mathbf{u}^1 \qquad (2.40)$$

where the transformation matrix $\mathbf{Q}$ is

$$\mathbf{Q} = \begin{bmatrix} \mathbf{I} \\ -\mathbf{K}_{22}^{-1}\mathbf{K}_{21} \end{bmatrix} \qquad (2.41)$$

and $\mathbf{I}$ is the unit (identity) matrix.

Substituting Equation 2.40 into Equation 2.38 and premultiplying the resulting equation by $\mathbf{Q}^{\mathrm{T}}$, one obtains a new reduced-order model

$$\mathbf{Q}^{\mathrm{T}}\mathbf{M}\mathbf{Q}\ddot{\mathbf{u}}^1 + \mathbf{Q}^{\mathrm{T}}\mathbf{K}\mathbf{Q}\mathbf{u}^1 = \mathbf{F} \qquad (2.42)$$

where

$$\mathbf{Q}^{\mathrm{T}}\mathbf{M}\mathbf{Q} = \mathbf{M}_{11} - \mathbf{M}_{12}\mathbf{K}_{22}^{-1}\mathbf{K}_{21} + \mathbf{K}_{21}^{\mathrm{T}}\mathbf{K}_{22}^{-1}\mathbf{M}_{22}\mathbf{K}_{22}^{-1}\mathbf{K}_{21} \qquad (2.43)$$

and

$$\mathbf{Q}^{\mathrm{T}}\mathbf{K}\mathbf{Q} = \mathbf{K}_{11} - \mathbf{K}_{12}\mathbf{K}_{22}^{-1}\mathbf{K}_{21} \qquad (2.44)$$

Newton's law

$$\mathbf{M}\ddot{u}(t) = -\mathbf{F}$$

Hooke's Law

$$\mathbf{F} = \mathbf{K}u(t)$$

Hamilton's principle

$$\delta \int_{t_1}^{t_2} (T - V)\mathrm{d}t + \int_{t_1}^{t_2} \delta W\,\mathrm{d}t = 0$$

Finite element equation without damping

$$\mathbf{M}\ddot{u} + \mathbf{K}u = \mathbf{F}$$

$$\mathbf{M} = \sum_e \mathbf{L}^{\mathrm{T}}\mathbf{M}^e\mathbf{L}, \qquad \mathbf{K} = \sum_e \mathbf{L}^{\mathrm{T}}\mathbf{K}^e\mathbf{L}, \qquad \mathbf{F} = \sum_e \mathbf{L}^{\mathrm{T}}\mathbf{F}^e$$

Guyan reduction scheme

$$\mathbf{Q}^{\mathrm{T}}\mathbf{M}\mathbf{Q}\ddot{u}^1 + \mathbf{Q}^{\mathrm{T}}\mathbf{K}\mathbf{Q}u^1 = \mathbf{F}$$

$$\mathbf{Q} = \begin{bmatrix} \mathbf{I} \\ -\mathbf{K}_{22}^{-1}\mathbf{K}_{21} \end{bmatrix}$$

2.3 Vibration Analysis

According to the vibration characteristics to be extracted, vibration analysis can be categorized into the following two types: natural vibration analysis, including modal analysis, and (forced) response analysis. Natural vibration analysis can extract natural vibration frequencies and the associated mode shapes, which is a matrix eigenvalue problem, and can result from a finite element discretization. The response analysis refers to the calculation of the response, which can be displacements, strain, or stress, when the system is subjected to time-varying excitation forces. The response analysis can be further divided into any one a combination of harmonic response analysis, transient response analysis, and response spectrum analysis, depending on the nature of excitation forces.

2.3.1 Natural Vibration

As noted in previous chapters, the natural vibration frequencies (or simply *natural frequencies*) and the associated mode shapes of a vibrating system are independent of excitation forces. In other words, they are intrinsic characteristics of the vibration problem. Therefore, they constitute an important part of vibration theory and vibration engineering. When vibration engineers specify design requirements in terms of vibration, they normally do so by restricting natural frequencies, and sometimes restricting mode shapes as well. For instance, in order to enhance the passenger comfort, vehicle designers have to ensure that the first few natural frequencies of the vehicle are not within a certain range; in order to avoid vibration resonance, the natural frequencies of a transmission shaft should be designed not to be identical or even close to the rotating speeds of the shaft; in order to effectively control vibration, vibration sensors and actuators have to be located at those places where the dominant mode shapes have large displacements.

From a mathematical standpoint, the natural vibration analysis of a multi-DoF system requires the solution of a matrix eigenvalue problem. According to the theory of second-order ordinary differential equations, the solution of Equation 2.34, when $\mathbf{F} = 0$, can be expressed as $\mathbf{u} = \mathbf{v}\,e^{i\omega t}$. By substituting $\mathbf{u} = \mathbf{v}\,e^{i\omega t}$ into Equation 2.34 and letting $\mathbf{F} = 0$, one can obtain

$$\omega^2 \mathbf{Mv} = \mathbf{Kv} \tag{2.45}$$

Equation 2.45 represents a generalized matrix eigenvalue problem. For an N-dimensional matrix pair $(\mathbf{M}, \mathbf{K})$, there exist N pairs of solutions $(\omega_i, \mathbf{v}_i)$, $\omega_i^2 \mathbf{Mv} = \mathbf{Kv}_i$ $(i = 1, 2, \ldots, N)$, where ω_i and $\mathbf{v}_i$ are called the ith natural frequency and the associated ith mode shape, respectively.

The numerical methods for solving the matrix eigenvalue problem given by Equation 2.45 have been well developed with the symmetric and sparse features of $(\mathbf{M}, \mathbf{K})$ being fully considered. Those that have been used by commercial finite element software packages include the power method, the subspace iteration method, the LR method, the QR method, the Givens method, the Householder method, and the Lanczos method.

When conducting vibration modeling, modelers need to understand how idealization and simplification will affect the resulting natural frequencies and the associated mode shapes. Idealization and simplification cause a difference between the actual mass matrix $\mathbf{M}$ and the resulting mass matrix $\mathbf{M}_r$ $(\mathbf{M} = \mathbf{M}_r + \Delta\mathbf{M})$, and a difference between the actual stiffness matrix $\mathbf{K}$ and the resulting stiffness matrix $\mathbf{K}_r$ $(\mathbf{K} = \mathbf{K}_r + \Delta\mathbf{K})$. Rayleigh's quotient [9,11] can be used to determine the effect of $\Delta\mathbf{K}$ and $\Delta\mathbf{M}$ on a particular natural frequency. Rayleigh's quotient is defined as

$$R(\mathbf{x}) = \frac{\mathbf{x}^T \mathbf{Kx}}{\mathbf{x}^T \mathbf{Mx}} \tag{2.46}$$

Note that Rayleigh's quotient $R(\mathbf{x})$ becomes the square of the ith natural vibration frequency, $R(\mathbf{x}) = \omega_i^2$, when $\mathbf{x} = \mathbf{v}_i$. Thus, Rayleigh's quotient can be expressed as

$$\omega_i^2 + \Delta\omega_i^2 = \frac{(\mathbf{v}_i^T + \Delta\mathbf{v}_i^T)(\mathbf{K}_r + \Delta\mathbf{K})(\mathbf{v}_i + \Delta\mathbf{v}_i)}{(\mathbf{v}_i^T + \Delta\mathbf{v}_i^T)(\mathbf{M}_r + \Delta\mathbf{M})(\mathbf{v}_i + \Delta\mathbf{v}_i)} \tag{2.47}$$

where $\Delta\omega_i^2$ and $\Delta\mathbf{v}_i$ are the increase of the ith natural frequency and the variation of the ith mode shape, respectively, induced by $\Delta\mathbf{K}$ and $\Delta\mathbf{M}$. Because of the fact that $R(\mathbf{x})$ reaches the stationary value when $\mathbf{x}$ is equal to the eigenvector $\mathbf{v}_i$, Equation 2.47 can be simplified as [1,4]

$$\Delta\omega_i^2 = \mathbf{v}_i^T(\Delta\mathbf{K} - \Delta\mathbf{M})\mathbf{v}_i \tag{2.48}$$

Equation 2.48 indicates that an increase in stiffness leads to a rise in a natural frequency, but an increase in mass causes a decrease in a natural frequency, as is intuitively clear.

2.3.2 Harmonic Response

Harmonic response analysis determines the response of a vibration system (model) to harmonic excitation forces. A typical output is a plot showing response (usually displacement of a certain DoF) versus frequency. This plot indicates how the response at a certain DoF, as a function of excitation frequency, changes with excitation frequency. The harmonic response can also be used to calculate the response to a general periodic excitation force, if it can be satisfactorily approximated by a summation of its major harmonic components.

Consider a harmonic excitation force, $\mathbf{F}(t) = \mathbf{F}_0\,e^{i\omega t}$. Substituting it into Equation 2.34, we have

$$\mathbf{M\ddot{u}}(t) + \mathbf{Ku}(t) = e^{i\omega t}\mathbf{F}_0 \tag{2.49}$$

According to the theory of differential equations, its steady solution can be written as $\mathbf{u}(t) = e^{i\omega t}\mathbf{U}$. After substitution of $\mathbf{u}(t) = e^{i\omega t}\mathbf{U}$ into Equation 2.49, one obtains

$$(-\omega^2\mathbf{M} + \mathbf{K})\mathbf{U} = \mathbf{F}_0 \tag{2.50}$$

Harmonic response analysis will solve Equation 2.50 for **U** against ω. There are many numerical methods available for solving Equation 2.50. The most efficient one is the modal superposition method.

In the modal superposition method, the response is expressed as a linear combination given by

$$\mathbf{u}(t) = \sum_{i=1}^{j} \mathbf{v}_i \widehat{\mathbf{u}}_i(t) = \mathbf{\Phi}\widehat{\mathbf{u}}(t) \tag{2.51}$$

where $\mathbf{\Phi} = [\mathbf{v}_1, \mathbf{v}_2, \ldots, \mathbf{v}_j]$ is a modal matrix that contains the dominant mode shapes, and $\widehat{\mathbf{u}}^{\mathrm{T}}(t) = [\widehat{\mathbf{u}}_1(t), \widehat{\mathbf{u}}_2(t), \ldots, \widehat{\mathbf{u}}_j(t)]$ are called modal coordinates. Substituting Equation 2.51 into Equation 2.49 and premultiplying the result by $\mathbf{\Phi}^{\mathrm{T}}$, we obtain

$$\ddot{\widehat{\mathbf{u}}}(t) + \mathbf{\Lambda}\widehat{\mathbf{u}}(t) = \mathbf{\Phi}^{\mathrm{T}}\, e^{i\omega t}\mathbf{F}_0 \tag{2.52}$$

Note that the modal matrix $\mathbf{\Phi}$ has already been normalized against the mass matrix ($\mathbf{\Phi}^{\mathrm{T}}\mathbf{M}\mathbf{\Phi} = \mathbf{I}$), and that $\mathbf{\Lambda} = \mathrm{diag}(\omega_1, \omega_2, \ldots, \omega_j)$.

Equation 2.52 represents a set of decoupled modal equations with a much smaller dimension than Equation 2.49. After solving Equation 2.52 for $\widehat{\mathbf{u}}(t)$ and transforming $\widehat{\mathbf{u}}(t)$ back to $\mathbf{u}(t)$ through Equation 2.51, we obtain $\mathbf{u}(t)$.

2.3.3 Transient Response

Transient response analysis (sometimes called time-history analysis) determines the dynamic response of a structure under the action of time-varying excitation. Excitation forces are explicitly defined in the time domain. The computed response usually includes the time-varying displacements, accelerations, strains, and stresses. Consider Equation 2.34 in its general form

$$\begin{cases} \mathbf{M}\ddot{\mathbf{u}}(t) + \mathbf{K}\mathbf{u}(t) = \mathbf{F}(t) \\ \mathbf{u}(0) = \mathbf{u}_0; \;\; \dot{\mathbf{u}}(0) = \dot{\mathbf{u}}_0 \end{cases} \tag{2.53}$$

where $\mathbf{F}(t)$ is the excitation force, $\mathbf{u}_0$ is the initial displacement, and $\dot{\mathbf{u}}_0$ is the initial velocity. As in the harmonic response analysis, Equation 2.53 can be solved by the modal superposition method.

Substituting Equation 2.51 into Equation 2.53, premultiplying the result of the first equation by $\mathbf{\Phi}^{\mathrm{T}}$, and premultiplying the result of the initial condition by $\mathbf{\Phi}^{\mathrm{T}}\mathbf{M}$, we obtain

$$\begin{cases} \ddot{\widehat{\mathbf{u}}}(t) + \mathbf{\Lambda}\widehat{\mathbf{u}}(t) = \mathbf{\Phi}^{\mathrm{T}}\mathbf{F}(t) \\ \widehat{\mathbf{u}}(0) = \mathbf{\Phi}^{\mathrm{T}}\mathbf{M}\mathbf{u}_0; \;\; \dot{\widehat{\mathbf{u}}}(0) = \mathbf{\Phi}^{\mathrm{T}}\mathbf{M}\dot{\mathbf{u}}_0 \end{cases} \tag{2.54}$$

Equation 2.54 represents a set of decoupled modal equations, which can be solved by means of numerical integration techniques. After solving Equation 2.54 for $\widehat{\mathbf{u}}(t)$ and transforming $\widehat{\mathbf{u}}(t)$ back to $\mathbf{u}(t)$ through Equation 2.51, we can obtain $\mathbf{u}(t)$.

To implement the numerical integration techniques, the overall time period being studied has to be divided into a number of smaller time steps. If the time step is too large, portions of the response (such as spikes) could be missed or truncated. On the other hand, if the time step is too small, the analysis will take an excessively long time or even a prohibitive amount of time.

2.3.4 Response Spectrum

The excitation forces, resulting from earthquakes, winds, ocean waves, jet engine thrust, uneven roads, and so on, do not have repeated patterns, for a variety of reasons, and thus it is difficult to describe them using a deterministic time history. Such excitations are normally treated as random excitations. The assumption that such excitation forces are random is recognition of our lack of knowledge of the detailed characteristics of the excitation forces. Some excitation forces, like those resulting from an uneven road, could be measured to any desired accuracy, and thus they would become deterministic rather than random. But it is not cost-effective and not convenient to do so. Therefore, engineers prefer to characterize these excitation forces by a statistical description that can be easily measured on any particular representative length of time history. Of the statistical descriptions, the autocorrelation function and the power spectral density function are the most important. Denote $f(t)$ as a stationary random excitation force, and $R_f(\tau)$ as its autocorrelation function and $S_f(\omega)$ as its power spectral density function. Their relations [6] are

$$R_f(\tau) = \lim \frac{1}{T} \int_0^T f(t) f(t+\tau) \, dt \tag{2.55}$$

$$S_f(\omega) = \int_{-\infty}^{\infty} R_f(\tau) e^{-i\omega\tau} \, d\tau \tag{2.56}$$

The response spectrum analysis here calculates the power spectral density function of the response of a vibration model to a random excitation force, described by its power spectral density function (see Chapter 5). For the single DoF system given in Figure 2.1

$$m\ddot{u}(t) + ku(t) = f(t) \tag{2.57}$$

The power spectral density function of the response $u(t)$ is given by

$$S_u(\omega) = |H(\omega)|^2 S_f(\omega) \tag{2.58}$$

where

$$H(\omega) = (-m\omega^2 + k)^{-1} \tag{2.59}$$

is the frequency response function representing the natural vibration characteristic of the system.

In the case of the multiple-DoF system given by Equation 2.34, the random excitation force $\mathbf{F}(t)$ is a column vector. For the sake of simplicity, we assume all of the components in the vector $\mathbf{F}(t)$ are stationary and statistically independent. Accordingly, all of the components in the vector of the response $\mathbf{u}(t)$ are stationary. Under this assumption, the power spectral density function of the response $\mathbf{u}(t)$ is determined by

$$\mathbf{S}_u(\omega) = \mathbf{H}(\omega) \mathbf{S}_f(\omega) \mathbf{H}^{\mathrm{T}}(\omega) \tag{2.60}$$

where $\mathbf{S}_f(\omega)$ is a diagonal matrix with its ith element as the power spectral density function of the ith element in $\mathbf{F(t)}$, and $\mathbf{H}(\omega)$ is the frequency response function matrix defined by

$$\mathbf{H}(\omega) = (-\mathbf{M}\omega^2 + \mathbf{K})^{-1} \tag{2.61}$$

From Equation 2.60 and Equation 2.61 one can see that the power spectral density function matrix of the response is correlated to the power spectral density function matrix of the excitation force by means of the frequency response matrix of the system $\mathbf{H}(\omega)$. In commercial finite element software packages, $\mathbf{H}(\omega)$ is often calculated by the truncated modal method in which $\mathbf{H}(\omega)$ is approximately expressed by the dominant natural frequencies and the associated mode shapes, neglecting the contributions of the other mode shapes to $\mathbf{H}(\omega)$, as given below.

$$\mathbf{H}(\omega) \approx \sum_i \frac{\mathbf{v}_i \mathbf{v}_i^{\mathrm{T}}}{\omega_i^2 - \omega^2} \tag{2.62}$$

Natural vibration analysis (generalized eigenvalue problem)

$$\omega^2 \mathbf{M} \mathbf{v} = \mathbf{K} \mathbf{v}$$

Rayleigh's quotient

$$R(\mathbf{x}) = \frac{\mathbf{x}^T \mathbf{K} \mathbf{x}}{\mathbf{x}^T \mathbf{M} \mathbf{x}}$$

Harmonic response

$$(-\omega^2 \mathbf{M} + \mathbf{K})\mathbf{U} = \mathbf{F}_0$$

$$\mathbf{u}(t) = \sum_{i=1}^{j} \mathbf{v}_i \widehat{\mathbf{u}}_i(t) = \mathbf{\Phi} \widehat{\mathbf{u}}(t)$$

$$\ddot{\widehat{\mathbf{u}}}(t) + \mathbf{\Lambda} \widehat{\mathbf{u}}(t) = \mathbf{\Phi}^T e^{i\omega t} \mathbf{F}_0$$

Transient response

$$\begin{cases} \mathbf{M}\ddot{\mathbf{u}}(t) + \mathbf{K}\mathbf{u}(t) = \mathbf{F}(t) \\ \mathbf{u}(0) = \mathbf{u}_0; \;\; \dot{\mathbf{u}}(0) = \dot{\mathbf{u}}_0 \end{cases}$$

$$\begin{cases} \ddot{\widehat{\mathbf{u}}}(t) + \mathbf{\Lambda} \widehat{\mathbf{u}}(t) = \mathbf{\Phi}^T \mathbf{F}(t) \\ \widehat{\mathbf{u}}(0) = \mathbf{\Phi}^T \mathbf{M} \mathbf{u}_0; \;\; \dot{\widehat{\mathbf{u}}}(0) = \mathbf{\Phi}^T \mathbf{M} \dot{\mathbf{u}}_0 \end{cases}$$

Response spectrum analysis

$$\mathbf{S}_u(\omega) = \mathbf{H}(\omega) \mathbf{S}_f(\omega) \mathbf{H}^T(\omega)$$

$$\mathbf{H}(\omega) = (-\mathbf{M}\omega^2 + \mathbf{K})^{-1}$$

2.4 Commercial Software Packages

There are many commercial finite element analysis (FEA) software packages available for vibration analysis, and they have been so well developed that they have an extensive range of vibration analysis capabilities. Some software packages are intended for generic engineering structures, for example, ABAQUS, ADINA, ALGOR, ANSYS, COSMOSWorks, MSC/NASTRAN, DYNA3D, and LS-DYNA. The others are designed for the vibration analysis of specific vibration problems. For example, the software package Bridge and the software package LUSAS Bridge are for the vibration analysis of bridges.

Normally, FEA modeling software has the following three major components: a preprocessor, a solver, and a postprocessor. The preprocessor is responsible for building up geometries, meshing, specification of element types, material properties, and boundary conditions; the solver solves the matrix equations; the postprocessor provides visualization of results and outputs the results in different formats. Because of the fact that those CAD/CAE systems such as CATIA, Unigraphics, Pro/E, and Solidworks have powerful capabilities for building up geometries, vibration modelers often import geometries from such systems rather than building them up using the FEA modeling software packages themselves.

In this section, we will select the software packages ABAQUS, ADINA, ALGOR, ANSYS, COSMOSWorks, MSC/NASTRAN, ABAQUS/Explicit, DYNA3D, and LS-DYNA, and will briefly introduce their major capabilities for vibration analysis. In fact, the vibration analysis capability is only a small portion of their total capabilities. All of these packages can perform basic vibration analysis including: (1) determination of natural mode shapes and frequencies, (2) transient response, (3) steady-state response resulting from harmonic loading, and (4) response spectrum analysis. In addition, each of them has its own strengths in some specific areas. These special capabilities are listed below.

2.4.1   ABAQUS

The ABAQUS software is for linear and nonlinear engineering analyses. Due to its wide range of functionality, ABAQUS usage spans many industries, including automotive, aerospace and defense, consumer electronics, manufacturing, medical, and rubber sealing.

Some special capabilities include

- Analysis of the coupled phenomena: thermo-mechanical, thermo-electrical, piezoelectric, pore fluid flow-mechanical, stress–mass diffusion, and shock and acoustic-structural
- Substructures and submodeling
- Material removal and addition
- Fracture mechanics design evaluation
- Parameterization and parametric studies
- User subroutines

2.4.2   ADINA

The ADINA system offers comprehensive finite element analyses of structures, fluids, and fluid flows with structural interactions. It is widely used in many fields, including the automotive, aerospace, manufacturing, nuclear, and biomedical industries, and in civil engineering and research.

Some special dynamic capabilities include

- Contact problems in statics and dynamics
- Substructuring in statics and dynamics
- Wave propagation and shock wave analysis

2.4.3   ALGOR

ALGOR's Professional Multiphysics includes capabilities for static structural analysis and Mechanical Event Simulation with linear and nonlinear material models, steady state and transient thermal analysis, electrostatic analysis, linear dynamics, and steady and unsteady fluid flow analysis. It is used in aerospace and space exploration; the automotive, transportation, consumer products, electronics, entertainment, manufacturing, chemical processing and medical industries; in the defense, power, and utility sectors; in civil engineering and scientific research; and in recreation and sports.

Some special features include

- Rigid-body motion
- Hertzian contact
- Submodeling
- Earthquake simulation
- Fluid–solid interaction
- The EAGLE programming language

2.4.4   ANSYS

ANSYS Software Suite offers capabilities for determining the structural, thermal, acoustic, electrostatic, magnetostatic, electromagnetic, and fluid-flow behavior of three-dimensional product designs, including the effects of multiphysics. The software simulates complex thermal/mechanical, fluid/structural and electrostatic/structural interactions. It is widely used in the aerospace, automotive, biomedical engineering, chemical engineering, civil engineering, communications, consumer products, defense, electronic packaging, industrial and scientific equipment production, and micro-electromechanical systems (MEMS) industries.

Some special dynamic capabilities include

- Modal analysis of prestressed structures
- Dynamic Topological design optimization
- Substructuring and submodeling
- Coupled field analysis of thermal-structural, fluid-structural, electrostatic-structural, magneto-structural, acoustic-structural, thermal-electric, thermal-electromagnetic, fluid-thermal, piezo-electric fields, and an electromechanical circuit simulator
- A parametric design language

2.4.5   COSMOSWorks

COSMOSWorks offers a wide spectrum of specialized analysis tools to virtually test and analyze complicated parts and assemblies, and is seamlessly integrated with Solidworks. COSMOSWorks is used for linear stress, strain, displacement, thermal analysis, design, optimization, and nonlinear analysis. Combined with ASTAR (Post Dynamics), COSMOSWorks is capable of more advanced dynamic analysis. It is used in a wide range of industries, including aerospace and defense, automotive and transportation, civil engineering, consumer products, electrical and electronics, heavy equipment, marine, medical and power.

The special dynamic features of the ASTAR module are

- Support of uniform and multi-base motion systems; the multi-base motion capability allows engineers to model structures with nonuniform support excitations.
- Support of the gap-friction element, which lets engineers model drop-test and other dynamic contact problems.
- The provision for several damper options such as Rayleigh damping, modal damping, concentrated damping, and composite modal damping.

2.4.6   MSC.Nastran

In 1965, MSC participated in a NASA-sponsored project to develop a unified approach to computerized structural analysis. The program became known as NASTRAN (NASA Structural Analysis Program); one of the first efforts to consolidate structural mechanics into a single computer program. The suite of MSC.Software is used in the space, aircraft, and automotive industries; in rail vehicle development; in general machinery; and in medical and electromechanical devices. Its capabilities include stress, vibration, heat transfer, acoustics, aeroelasticity, and coupled system analysis.

MSC.NASTRAN's special dynamic capabilities include

- Damping
- Direct matrix input
- Dynamic equations of motion
- Residual vector methods
- Enforced motion
- Complex eigenvalue analysis
- Normal mode of preloaded structures
- Dynamic design optimization
- Test-analysis correlation

2.4.7 ABAQUS/Explicit

ABAQUS/Explicit uses explicit time integration for time stepping and addresses the following special types of analysis:

- Explicit dynamic response with or without adiabatic heating effects
- Fully coupled transient dynamic temperature–displacement procedure; explicit algorithms are used for both the mechanical and thermal response
- Annealing for multistep forming simulations
- Acoustic and coupled acoustic-structural analysis
- Automatic adaptive meshing, which allows robust solutions of highly nonlinear problems

2.4.8 DYNA3D

DYNA3D is an explicit finite element program for structural and continuum mechanics problems. Due to its explicit nature, DYNA3D uses small time steps to integrate the equations of motion and is especially efficient at solving transient dynamic problems.

The specific analysis capabilities of DYNA3D include

- Static analysis using dynamic relaxation
- Dynamic analysis with static initialization from a NIKE3D implicit analysis
- Various contact slideline options for different contact situations between two bodies

2.4.9 LS-DYNA

LS-DYNA is a general-purpose transient dynamic finite element program, and is suited for complex dynamics, vibration, and wave propagation problems. Its explicit algorithm can be used for high-speed impact, shock, and vibration problems. Falling impact, rubber elasticity, and impact on sports goods (rackets, bats, and helmets) are typical examples of problems that can be handled by LS-DYNA. It is widely used in earthquake engineering; crashworthiness and occupant safety analysis; metal forming; biomedical engineering; train crashworthiness testing; sports, airbag, and seat-belt deployment; and in military, manufacturing, metal cutting, and bird strike applications.

The special capabilities of LS-DYNA include

- FEM-rigid multi-body dynamics coupling
- Underwater shock analysis
- Failure analysis
- Crack propagation analysis
- Real-time acoustics
- Design optimization
- Implicit springback analysis
- Multi-physics coupling
- Adaptive re-meshing
- Smooth particle hydrodynamics
- The element-free meshless method

2.5 The Basic Procedure of Vibration Analysis

In this section, a typical procedure in using commercial software packages to conduct vibration analysis is outlined.

2.5.1 Planning

This is a very important part of the entire analysis process, as it helps to ensure the success of the modeling. The quality of the results is strongly dependent on how accurately the model represents the actual problem being investigated. In order to generate a representative finite element model, all influencing factors must be scrutinized to determine whether their effects are considerable or negligible in the final result. The aspects listed below should be given consideration in the planning stage.

- *Modeling objectives.* Why is the vibration analysis required? What is the major concern of designers? What are the working conditions? Does the FEA model have to be used for static stress analysis as well as vibration analysis? These considerations affect how the FEA model is to be built up.
- *Modeling considerations.* Which type of analysis is required: natural vibration analysis or response analysis? What types of elements should be used? Where are loads and constraints applied? Can the model reduction/simplification resulting from symmetrical geometries and loading conditions be applied? There are no universal guidelines for these, but the aspects below can help you to make decision:
  - If the stress varies linearly through the thickness of thin-walled regions, shell elements can be used. If it varies parabolically, then at least three solid, second-order elements are required through the thickness in order to resolve a representative state of stress.
  - If a frequency or buckling analysis is being conducted, a full three-dimensional analysis may be needed to identify non-symmetric mode shapes.
  - If the region of interest is local, then a submodel may be appropriate, as it will save considerable time achieving a solution.
  - Large gradients in stress levels will require a high mesh density to capture the behavior appropriately.
  - The effects of simplifications on boundary conditions should be well predicted, for example: some over-constrained boundary conditions may result in higher natural frequencies of the finite element model.

The degree of accuracy of a model is very much dependant on the level of planning that has been carried out. Careful planning is the key to a successful analysis.

2.5.2 Preprocessing

The preprocessor stage in a general FEA package involves the following:

- *Defining the element type as planned before the analysis.* This may be one-, two-, or three-dimensional.
- *Creating the geometry.* The geometry is drawn in one-, two-, or three-dimensional space according to what kind of elements are going to be used. The model may be created in the preprocessor, or it can be imported from other CAD or CAE systems *via* a neutral file format (IGES, STEP, ACIS, Parasolid, DXF, etc.). The same units should be applied in all models, otherwise the results will be difficult to interpret or, in extreme cases, the results will not show up mistakes made during the loading and restraining of the model.
- *Applying a mesh.* Mesh generation is the process of dividing the continuum into a number of discrete parts or finite elements. The finer the mesh, the more accurate the result, but the longer the processing time. Therefore, a compromise between accuracy and solution speed is usually made. The mesh may be created manually or generated automatically, or, as in most cases, in a combined manner.

  Manual meshing is a long and tedious process for models with a fair degree of geometric complication, but with useful tools emerging in preprocessors, the task is becoming easier. Automatic mesh generators are very useful and popular. The mesh is created automatically by a mesh engine; the only requirement is to define the mesh density along the model's edges.

Automatic meshing has limitations as regards mesh quality and solution accuracy. Automatic brick element meshers are limited in function, but are steadily improving. Any mesh is usually applied to the model by simply selecting the mesh command on the preprocessor list of the user interface.

Usually a complex geometry needs to be decomposed into many smaller components in order to use the automatic meshing tool.

- *Assigning properties.* Material properties (Young's modulus; Poisson's ratio; the density; and if applicable, the coefficients of expansion, friction, thermal conductivity, damping effect, specific heat, etc.) will have to be defined. In addition element properties may need to be set. If two-dimensional elements are being used, the thickness property is required. One-dimensional beam elements require area, $I_{xx}, I_{yy}, I_{zz}, J$, and the direction of the beam axis in three-dimensional space. Shell elements, which are two-dimensional elements in three-dimensional space, require orientation and neutral surface offset parameters to be defined. Special elements such as mass, contact, spring, gap, coupling, damper, and so on require properties (specific to the element type) to be defined for their use.

- *Applying loads.* In the case of transient response analysis, some type of load is usually applied to the analysis model. The loading may be in the form of a point force, a pressure or a displacement, or a temperature or heat flux in a thermal analysis. The loads may be applied to a point, an edge, a surface, or even a complete body. The loads should be in the same unit system as the model geometry and material properties specified. In the case of modal analyses, a load does not have to be specified for the analysis to run.

- *Applying boundary conditions.* Structural boundary conditions are usually in the form of zero displacements; thermal boundary conditions are usually specified temperatures. A boundary condition may be specified to act in all directions (x, y, z), or in certain directions only. Boundary conditions can be placed on nodes, key points, lines, or areas. The boundary conditions applied on lines or areas can be of a symmetric or antisymmetric type, one allowing inplane rotations and out of plane translations, the other allowing in plane translations and out of plane rotations for a given line or area. The application of correct boundary conditions is critical to the accurate solution of the design problem.

2.5.3   Solution

The FEA solver can be logically divided into three main parts: the presolver, the mathematical-engine, and the postsolver. The presolver reads in the model created by the preprocessor and formulates the mathematical representation of the model. All the parameters defined during the preprocessing stage are used to do this, so if something has been omitted the presolver is very likely to stop the call to the mathematical-engine. If the model is correct, the solver proceeds to form the element stiffness matrix and the element mass matrix for the problem and calls the mathematical-engine, which calculates the result. The results are returned to the solver and the postsolver is used to calculate strains, stresses, velocities, response, and so on for each node within the component or continuum. All these results are sent to a result file that may be read by the postprocessor.

2.5.4   Postprocessing

Here the results of the analysis are read and interpreted. They can be presented in the form of a table, a contour plot, a deformed shape of the component, or the mode shapes and natural frequencies if frequency analysis is involved. Most postprocessors provide animation tools.

Contour plots are usually the most effective way of viewing results for structural type problems. Slices can be made through three-dimensional models to facilitate the viewing of internal stress and deformation patterns.

All postprocessors now include the calculation of stresses and strains in any of the x, y, or z directions; or indeed in a direction at an angle to the coordinate axes. The principal stresses and strains may also be plotted, or if required, the yield stresses and strains according to the main theories of failure (Von Mises, St Venant, Tresca, etc.).

2.5.5 Engineering Judgment

For many reasons, the vibration analysis results may not represent the actual vibration problem very well. Software packages will not reveal anything about this, and so it is the responsibility of modelers to make judgments.

Sound judgment comes from a thorough understanding of the actual vibration problem; indepth knowledge of vibration theory, FEA, and the software package used; and also rich experience in modeling. When you are not confident of your vibration analysis results you should check the following:

1. What units have been used, SI units or Imperial units? Are the units used consistent and compatible with the software package you are using?
2. What are the material's properties?
3. Is the boundary free or partially constrained, or flexibly connected to other parts?
4. How are the interconnections between different parts modeled (e.g., the interconnection between a two-dimensional plate and a three-dimensional block)?

Sometimes, a judgment is made by comparing your model's results and the results of different models that are similar in some sense to the one of concern, but which have been validated. A judgment can also be made by vibration measurements or testing under laboratory conditions or in real-life situations. By properly exploiting the combined test and analysis data, modelers can effectively and reliably identify otherwise only approximately known structural properties (e.g., joint stiffness), material properties, and loading; validate and refine the FEA model (simplification validation, model updating, etc.) by using test results as reference data; identify unknown or badly known physical properties; and better assess uncertainties in the FEA model.

2.6 An Engineering Case Study

In this section, we illustrate the procedure for the vibration analysis of a gearbox housing, shown in Figure 2.3. The vibration analysis was performed using ANSYS [5].

2.6.1 Objectives

The chief aim of the vibration analysis is to ensure that the gearbox housing is not subject to a dangerous resonant condition during the full range of operation. Specifically, the natural vibration frequencies of the gearbox housing have to be widely separated from the rotating speeds of the shafts. Hence, natural vibration analysis is required for this purpose. Furthermore, there are concerns about the strength of the gearbox, and so a static stress analysis is also required.

2.6.2 Modeling Strategy

The gearbox housing shown in Figure 2.3 contains the following three subparts: the vertical cylinder, the front housing, and the rear housing, which are welded together. Because the FEA model has to be built up with the considerations of both vibration analysis and static stress analysis, some detailed

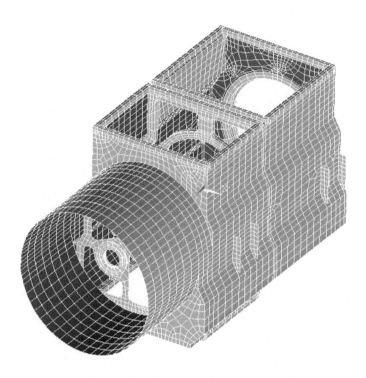

FIGURE 2.3   Finite element model of the gearbox housing. (Courtesy of Pacific Rim Engineered Products, Surrey, British Columbia.)

geometries such as filets are not deleted but modeled with fine meshes. In addition, a sufficient level of attention is paid to the interconnections between different sections. To achieve a balance between the accuracy of the results and the size of finite element model, quadratic elements (midsize nodes) are used for both shell elements and solid elements. These usually yield better results at less expense than linear elements.

Three types of finite elements are used to model the different parts:

1. The front cylinder plate, side plates and bottom plate: shell elements with variable thickness; 8 nodes, 6 DoF per node.
2. The ribs and fringes: solid elements; 20 nodes, 3 DoF per node.
3. The gears, shafts, clutch, and bearings: lumped mass elements; 1 node, 6 DoF per node.

Because of the discrepancy in the DoF between the shell elements and the solid elements, the nodal rotating freedoms around the edges that connect shell elements and solid elements together are not constrained, and consequently each rotating freedom needs to be constrained by two nodal translation freedoms on solid elements near the edge, but not on the edge.

The total numbers of nodes, shell elements, and solid elements are given in Table 2.1, and the complete finite element model is shown in Figure 2.3.

TABLE 2.1   Total Size of the Finite Element Model of the Gearbox Housing

Nodes	36,523
Shell elements	4,060
Solid elements	3,760

TABLE 2.2   Material Properties

Density	7800 kg/m^3
Young's modulus	2.1×10^{11} Pa
Poisson coefficient	0.29

TABLE 2.3   The First 10 Natural Frequencies

No.	Frequency (Hz)
1	46.46
2	67.73
3	81.57
4	105.5
5	166.2
6	204.6
7	205.1
8	212.0
9	213.4
10	222.8

2.6.3   Boundary Conditions

The four mountings on each side are constrained completely and the front edge of the cylinder is also completely constrained.

2.6.4   Material

The mechanical properties of the material are given in Table 2.2.

2.6.5   Results

The first 10 natural frequencies and the associated mode shapes are calculated with a Lanczos algorithm. They are listed in Table 2.3.

The first and the fifth mode shapes are shown in Figure 2.4 and Figure 2.5, respectively.

2.7   Comments

Vibration modeling using the finite element method is extremely powerful. However, with comforting contour plots, one can be easily deceived into thinking that a superior result has been achieved. Nevertheless, the quality of the result directly depends upon how accurately the model represents the actual physical problem being investigated. This involves three things: sufficient understanding of the actual vibration problem, sufficient knowledge of vibration theory including FEA, and hands-on experience in running an FEA software package. In particular, modelers have to understand the limitations of the theories applied and the numerical methods used. For example, the FEA can predict global characteristics such as natural frequencies of vibration and mode shapes more accurately than localized features such as stresses. This is an intrinsic nature of finite element methods. Without knowing this, modelers might incorrectly use an unnecessarily fine mesh for mode shape analysis while applying coarse meshes to evaluate stress.

FIGURE 2.4   The first mode shape (46.5 Hz) of the gearbox housing. (Courtesy of Pacific Rim Engineered Products, Surrey, British Columbia.)

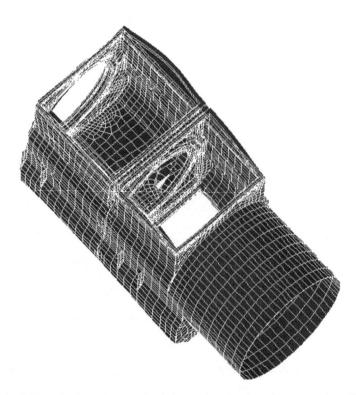

FIGURE 2.5   The fifth mode shape (166.2 Hz) of the gearbox housing. (Courtesy of Pacific Rim Engineered Products, Surrey, British Columbia.)

References

1. Chen, S. 1993. *Matrix Perturbation Theory in Structural Dynamics*, International Academic Publishers, Beijing.
2. den Hartog, J.P. 1984. *Mechanical Vibrations*, 4th ed., McGraw-Hill, New York.
3. Guyan, R.I., Reduction of stiffness and mass matrices, *AIAA Journal*, 3, 2, 380–381, 1965.
4. Hu, H.C. 1984. *Variational Principles of Theory of Elasticity with Applications*, Science Press, Beijing.
5. Moaveni, S. 1999. *Finite Element Analysis. Theory and Application with ANSYS*, Prentice Hall, Upper Saddle River, NJ.
6. Newland, D.E. 1993. *An Introduction to Random Vibration, Spectral and Wavelet Analysis*, Longman Scientific & Technical, New York.
7. Petyt, M. 1990. *Introduction to Finite Element Vibration Analysis*, Cambridge University Press, UK.
8. Rao, S.S. 1995. *Mechanical Vibrations*, 3rd ed., Addison-Wesley, Reading, MA.
9. Rayleigh, J.W.S. 1945. *Theory of Sound*, Vol. 1/2, Dover Publications, New York.
10. Turner, M.J., Clough, R.W., Martin, H.C., and Topp, L.J., Stiffness and deflection analysis of complex structures, *J. Aeronaut. Sci.*, 23, 805–823, 1956.
11. Washizu, K. 1982. *Variational Methods in Elasticity & Plasticity*, 3rd ed., Pergamon Press, UK.
12. Zienkiewicz, O.C. 1987. *The Finite Element Method*, 4th ed., McGraw-Hill, London.

3

Computer Analysis of Flexibly Supported Multibody Systems

Ibrahim Esat
Brunel University

M. Dabestani
Furlong Research Foundation

Summary

This chapter presents the Euler–Newton formulation of oscillatory behavior of a multibody system interconnected by discrete stiffness elements. Bodies are interconnected by springs, and/or dashpots (dampers). Connections are described in terms of end coordinates of springs relative to the coordinate system of the body to which it is attached. Stiffness characteristics are described along the three principal axes of springs. Orientation of springs and masses are described by using appropriate Euler angles. The model developed is linear, and gyroscopic influences are ignored. The chapter gives a detailed treatment of rigid bodies in three dimensional space using vector-matrix formulation. Complete formulation and assembly issues relating to programming aspects are presented. A software suite called VIBRATIO, based on the present formulation, is described. The capabilities of VIBRATIO are indicated and illustrative examples are given in both frequency and time domains. A student version of VIBRATIO is available at no cost to the users of this book at www.signal-research.com.

3.1 Introduction

There are many commercial software packages for analysis of kinematics and dynamics of multibody linkage systems. There are fewer software tools for analysis of vibration of multiple rigid-body systems in 3-D space, even though some finite element analysis (FEA) packages offer rigid-body capability. Common FEA software packages treat rigid bodies using point masses or point inertias. Although this is not a serious restriction, when it comes to attaching discrete stiffness elements to a body away from its center of gravity (COG), the attachment is achieved by introducing "lever arms" with a very high Young's modulus. One may argue that the error introduced in doing so is

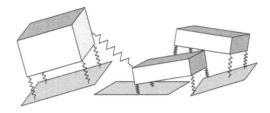

FIGURE 3.1 Schematic representation of a multibody system.

acceptable but how true this argument is depends on the problem, and there is no escape from the fact that this approach can create ill-conditioned stiffness matrices. The correct way, however, is to incorporate the created kinematic constraints into rigid-body geometries. This is the approach presented in this chapter. A typical rigid multibody system supported or interconnected by discrete spring elements, as considered here, is shown in Figure 3.1. The chapter presents a complete formulation of a multibody system flexibly supported by linear mountings. The formulations and methods proposed in this chapter are used in the VIBRATIO suite of vibration analysis software.

3.2 Theory

3.2.1 Definitions and Assumptions

- Springs have zero length.
- The stiffness parameters of the springs in their principal axes of deflection remain uncoupled.
- The amplitude of oscillation is small. No geometrical nonlinearity is involved. In other words, the orientation of both mountings and bodies remains unaffected by oscillations.
- The time-dependent effects of polymeric material are excluded.
- Gyroscopic effects are negligible.

These assumptions are acceptable for most engineering vibration problems with small amplitude vibration.

3.2.2 Equations of Motion for the Linear Model

To set up equations of motion for a dynamic system, the following steps are required:

(i) Generation of the equations of internal reactions and external forces. The internal reactions due to damping and stiffness elements have to be expressed in a unified and structured fashion for formulation of the stiffness matrix. (The damping matrix structure is identical to the stiffness matrix structure, except that stiffness coefficients need to be replaced by damping coefficients.)
(ii) Generation of the equations of linear momentum (force–acceleration equations).
(iii) Generation of the equations of angular momentum (turning moment equations).

3.2.3 Linear Momentum–Force Systems

3.2.3.1 Stiffness and Damping Systems

The formulations applied in this chapter to obtain the stiffness matrix apply equally to the damping matrix by replacing stiffness parameters with their corresponding damping parameters.

Let us assume that spring stiffness parameters are described in a local three-dimensional (3-D) Cartesian coordinate frame, the axes of which coincide with the principal axes of the springs. The force vector $\mathbf{f}$ acting on the springs may be expressed as

$$\mathbf{f} = \mathbf{kx} \tag{3.1}$$

where $\mathbf{k}$ is the stiffness matrix (diagonal with principal stiffness values) and $\mathbf{x}$ is the displacement vector (expressing the spring extension).

In general, it is convenient to describe the behavior of a system in the global coordinate frame, OXYZ. This is not a prerequisite for the formulation. It is equally possible to obtain equations of motion for each body in its own frame. In this chapter, all spring stiffness matrices will be expressed in a common global coordinate frame. The individual spring matrices will be transformed accordingly. Since the principal axes of the springs and the global coordinates are all orthogonal, an orthogonal transformation exists between the two frames. A vector, $\mathbf{x}$, in the local coordinates could be expressed as a vector, $\mathbf{X}$, in the global coordinate system. Using, $\mathbf{T}$, a transformation matrix

$$\mathbf{X} = \mathbf{Tx} \tag{3.2}$$

If we premultiply Equation 3.1 by $\mathbf{T}$, then we have $\mathbf{Tf} = \mathbf{Tkx}$. But $\mathbf{Tf} = \mathbf{F}$.

Therefore, force vector, $\mathbf{F}$, in the global coordinate frame, may be written as

$$\mathbf{F} = \mathbf{Tkx} \tag{3.3}$$

For consistency, $\mathbf{x}$ needs to be replaced by $\mathbf{X}$. To replace $\mathbf{x}$ by $\mathbf{X}$, Equation 3.2 may be used, giving $\mathbf{x} = \mathbf{T}^T\mathbf{X}$. This is true since $\mathbf{T}^{-1} = \mathbf{T}^T$ for orthogonal transformation matrices. Therefore,

$$\mathbf{F} = \mathbf{TkT}^T\mathbf{X} \tag{3.4}$$

Then introduce a new matrix $\mathbf{K}$, where

$$\mathbf{K} = \mathbf{TkT}^T$$

Now $\mathbf{TkT}^T$ is the stiffness matrix of the spring in the global coordinate system. The transformation matrix $\mathbf{T}$ may be described in three Euler angles of rotation.

3.2.3.2 Generalization of the Equation of Linear Momentum

If the mass/inertia matrix in the Euler–Newton formulation is obtained relative to the axes passing through the center of mass, then the submatrix of the mass matrix corresponding to linear momentum is a diagonal matrix containing the mass elements; thus,

$$\mathbf{h}_i = \mathbf{m}_i\mathbf{v} \tag{3.5}$$

Here, $\mathbf{h}_i$ is linear momentum, $\mathbf{m}_i$ is a diagonal matrix, and $\mathbf{v}$ is the velocity vector of the center of mass (casually known as COG) of the body.

The usual transformation to the global coordinate frame, $\mathbf{H}_l = \mathbf{TmT}^T\mathbf{v}$, leaves the mass matrix, $\mathbf{m}$, unchanged. Therefore, the force acting on a body, i (i.e., the rate of change of linear momentum), may be expressed simply as

$$\text{Force} = \dot{\mathbf{H}}_l = \frac{\partial \mathbf{H}_l}{\partial t} = \mathbf{ma} \tag{3.6}$$

where $\mathbf{a}$ is the acceleration vector of the COG.

3.2.4 Generalization of the Equations of Moment of Momentum

The equations of moment of momentum may be expressed as

$$\mathbf{h}_a = \mathbf{j\omega} \tag{3.7}$$

where $\mathbf{h}_a$ is the angular momentum vector, $\mathbf{j}$ is the moments of inertia matrix and $\boldsymbol{\omega}$ is the angular velocity of the coordinate frame. (In this case, the frame is attached to the body.)

Here, **j** may or may not be a diagonal matrix. However, it is always symmetric. Equation 3.7 is described in the local coordinate system of the rigid body and it has to be expressed in the global coordinate system for the final matrix assembly. As presented for the stiffness elements, transformation follows exactly the same steps as before. In this case, **T** refers to the transformation matrix of mass relative to the global coordinate system. Transforming Equation 3.7 to the global coordinates, we get

$$\mathbf{H}_a = \mathbf{T}\mathbf{j}\mathbf{T}^\mathrm{T}\Omega \tag{3.8}$$

Introduce a new matrix notation

$$\mathbf{J} = \mathbf{T}\mathbf{j}\mathbf{T}^\mathrm{T} \tag{3.9}$$

The vector differentiation of $\mathbf{H}_a$ gives the moment vector in the global coordinates

$$M = \dot{\mathbf{H}}_a = \frac{\partial \mathbf{H}_a}{\partial t} + \boldsymbol{\omega} \times \mathbf{H}_a \tag{3.10}$$

where $\boldsymbol{\omega}$ is the angular velocity of the body (or the coordinate frame, as the body is fixed to the frame).

Note that $\boldsymbol{\omega} \times \mathbf{H}_a$ contains the product of angular velocity terms and this, for small and geometrically linear vibration problems, is small and may be ignored.

3.2.5 Assembly of Equations

To assemble the equations of motion, the internal forces acting on individual bodies due to their motion relative to each other are required. In Figure 3.2, two bodies (**i** and **j**) in motion are shown, connected by spring $\mathbf{K}_p$.

Motion of the origin of vector **i** (which coincides with the COG of body i) is given by $\mathbf{x}_i = (x_i, y_i, z_i)$, and the angular rotation of the coordinates is given by $\boldsymbol{\alpha}_I = (\alpha_i, \beta_i, \gamma_i)$. Similarly, the motion of body j is described by $\mathbf{x}_j = (x_j, y_j, z_j)$ and $\boldsymbol{\alpha}_j = (\alpha_j, \beta_j, \gamma_j)$.

For small motions, displacements of the end points of the springs on each body, described in the *coordinate frame* of each body, are given by

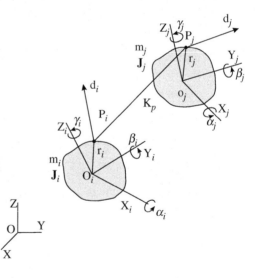

FIGURE 3.2 Two bodies connected by springs.

$$\mathbf{d}_i = \mathbf{x}_i + \boldsymbol{\alpha}_i \times \mathbf{r}_{pi} \tag{3.11}$$

$$\mathbf{d}_j = \mathbf{x}_j + \boldsymbol{\alpha}_j \times \mathbf{r}_{pj} \tag{3.12}$$

where $\mathbf{r}_{pj}$ and $\mathbf{r}_{pj}$ are the coordinates of the spring attachment relative to the bodies *i* and *j* in their respective coordinate frames, given as $\mathbf{r}_{pi} = (x_{pi}, y_{pi}, z_{pi})$ and $\mathbf{r}_{pj} = (x_{pj}, y_{pj}, z_{pj})$.

Cross-product terms in Equation 3.11 and Equation 3.12 can be converted into matrix form as

$$\boldsymbol{\alpha}_i \times \mathbf{r}_{pi} = \begin{bmatrix} 0 & z_{pi} & -y_{pi} \\ -z_{pi} & 0 & x_{pi} \\ y_{pi} & -x_{pi} & 0 \end{bmatrix} \begin{Bmatrix} \alpha_i \\ \beta_i \\ \gamma_i \end{Bmatrix} \tag{3.13}$$

and

$$\alpha_j \times \mathbf{r}_{pi} = \begin{bmatrix} 0 & z_{pj} & -y_{pj} \\ -z_{pj} & 0 & x_{pj} \\ y_{pj} & -x_{pj} & 0 \end{bmatrix} \begin{Bmatrix} \alpha_j \\ \beta_j \\ \gamma_j \end{Bmatrix} \tag{3.14}$$

Let us choose the matrix notation $\mathbf{R}_{pi}$ as

$$\mathbf{R}_{pi} = \begin{bmatrix} 0 & z_{pi} & -y_{pi} \\ -z_{pi} & 0 & x_{pi} \\ y_{pi} & -x_{pi} & 0 \end{bmatrix} \tag{3.15}$$

and $\mathbf{R}_{pj}$ as

$$\mathbf{R}_{pj} = \begin{bmatrix} 0 & z_{pj} & -y_{pj} \\ -z_{pj} & 0 & x_{pj} \\ y_{pj} & -x_{pj} & 0 \end{bmatrix} \tag{3.16}$$

Therefore,

$$\mathbf{d}_i = \mathbf{x}_i + \alpha_i \times \mathbf{r}_{pi}$$

which is

$$= \begin{Bmatrix} x_i \\ y_i \\ x_i \end{Bmatrix} + \begin{bmatrix} 0 & z_{pi} & -y_{pi} \\ -z_{pi} & 0 & x_{pi} \\ y_{pi} & -x_{pi} & 0 \end{bmatrix} \begin{Bmatrix} \alpha_i \\ \beta_i \\ \gamma_i \end{Bmatrix} \tag{3.17}$$

Using the new notation, we have

$$\mathbf{d}_i = \mathbf{x}_i + \mathbf{R}_{pi}\alpha_i \tag{3.18}$$

Now $\mathbf{d}_j$ is given by

$$\mathbf{d}_j = \mathbf{x}_j + \alpha_j \times \mathbf{r}_{pj} \tag{3.19}$$

and can be written as

$$\mathbf{d}_j = \begin{Bmatrix} x_j \\ y_j \\ x_j \end{Bmatrix} + \begin{bmatrix} 0 & z_{pj} & -y_{pj} \\ -z_{pj} & 0 & x_{pj} \\ y_{pj} & -x_{pj} & 0 \end{bmatrix} \begin{Bmatrix} \alpha_j \\ \beta_j \\ \gamma_j \end{Bmatrix}$$

and therefore in matrix notation, we have

$$\mathbf{d}_j = \mathbf{x}_j + \mathbf{R}_{pj}\alpha_j \tag{3.20}$$

To calculate the reactions acting on each body, the relative displacements between the connecting points (stretch) should be calculated.

The relative displacements are given by

$$\mathbf{d} = \mathbf{d}_j - \mathbf{d}_i \tag{3.21}$$

The reaction forces due to the relative displacements on each body are given by

$$\mathbf{Fs}_i = \mathbf{K}_p\mathbf{d} \tag{3.22}$$

For equal but opposite directions, we have

$$\mathbf{Fs}_j = -\mathbf{K}_p\mathbf{d} \tag{3.23}$$

Moments for spring forces acting at points $\mathbf{r}_i$ and $\mathbf{r}_j$ on bodies i and j, respectively, are given by

$$\mathbf{M}_i = \mathbf{r}_i \times \mathbf{Fs}_i \tag{3.24}$$

On body j, we have

$$\mathbf{M}_j = \mathbf{r}_j \times \mathbf{Fs}_j \tag{3.25}$$

The cross-products may be expressed in matrix form as

$$\mathbf{M}_i = \mathbf{r}_i \times \mathbf{Fs}_i = \begin{bmatrix} 0 & -z_{pi} & y_{pi} \\ z_{pi} & 0 & -x_{pi} \\ -y_{pi} & x_{pi} & 0 \end{bmatrix} \begin{Bmatrix} Fs_{xi} \\ Fs_{yi} \\ Fs_{zi} \end{Bmatrix} \tag{3.26}$$

$$\mathbf{M}_j = \mathbf{r}_j \times \mathbf{Fs}_j = \begin{bmatrix} 0 & -z_{pj} & y_{pj} \\ z_{pj} & 0 & -x_{pj} \\ -y_{pj} & x_{pj} & 0 \end{bmatrix} \begin{Bmatrix} Fs_{xj} \\ Fs_{yj} \\ Fs_{zj} \end{Bmatrix} \tag{3.27}$$

Note that the matrices in Equation 3.26 and Equation 3.27 are transposed forms of the matrices in Equation 3.15 and Equation 3.16

$$\mathbf{R}_{pi}^{\mathrm{T}} = \begin{bmatrix} 0 & -z_{pi} & y_{pi} \\ z_{pi} & 0 & -x_{pi} \\ -y_{pi} & x_{pi} & 0 \end{bmatrix} \tag{3.28}$$

$$\mathbf{R}_{pj}^{\mathrm{T}} = \begin{bmatrix} 0 & -z_{pj} & y_{pj} \\ z_{pj} & 0 & -x_{pj} \\ -y_{pj} & x_{pj} & 0 \end{bmatrix} \tag{3.29}$$

Now the equations of motion can be compiled for the translation of body i

$$m_i \ddot{\mathbf{x}}_i + \mathbf{K}_p \mathbf{d}_i - \mathbf{K}_p \mathbf{d}_j = \mathbf{F}_i \tag{3.30}$$

In this case, $\mathbf{F}_i$ is the vector of external forces acting on body i. Substituting $\mathbf{d}_i$ and $\mathbf{d}_j$ into Equation 3.30, from Equation 3.18 and Equation 3.20 we have

$$m_i \ddot{\mathbf{x}}_i + \mathbf{K}_p(\mathbf{x}_i + \mathbf{R}_{pi}\boldsymbol{\alpha}_i) - \mathbf{K}_p(\mathbf{x}_j + \mathbf{R}_{pj}\boldsymbol{\alpha}_j) = \mathbf{F}_i \tag{3.31}$$

Expanding this, we get

$$m_i \ddot{\mathbf{x}}_i + \mathbf{K}_p \mathbf{x}_i + \mathbf{K}_p \mathbf{R}_{pi}\boldsymbol{\alpha}_i - \mathbf{K}_p \mathbf{x}_j - \mathbf{K}_p \mathbf{R}_{pj}\boldsymbol{\alpha}_j = \mathbf{F}_i \tag{3.32}$$

Similarly, for body j, substituting the expressions for $\mathbf{d}_i$ and $\mathbf{d}_j$, we get

$$m_j \ddot{\mathbf{x}}_j + \mathbf{K}_p \mathbf{d}_i - \mathbf{K}_p \mathbf{d}_j = \mathbf{F}_j \tag{3.33}$$

Again, $\mathbf{F}_j$ in this case is the vector of external forces acting on body j.

$$m_j \ddot{\mathbf{x}}_j + \mathbf{K}_p(\mathbf{x}_i + \mathbf{R}_{pi}\boldsymbol{\alpha}_i) - \mathbf{K}_p(\mathbf{x}_j + \mathbf{R}_{pj}\boldsymbol{\alpha}_j) = \mathbf{F}_j \tag{3.34}$$

$$m_j \ddot{\mathbf{x}}_j + \mathbf{K}_p \mathbf{x}_i + \mathbf{K}_p \mathbf{R}_{pi}\boldsymbol{\alpha}_i - \mathbf{K}_p \mathbf{x}_j - \mathbf{K}_p \mathbf{R}_{pj}\boldsymbol{\alpha}_j = \mathbf{F}_j \tag{3.35}$$

With Equation 3.32 and Equation 3.35, the force–acceleration equations are complete.

Moment Equations

The moment equation may be written for body i as shown in Equation 3.36, where $\mathbf{M}_i$ is the external moment acting on body i.

$$J_i \ddot{\boldsymbol{\alpha}}_i + \mathbf{r}_i \times (\mathbf{K}_p \mathbf{d}_i - \mathbf{K}_p \mathbf{d}_j) = \mathbf{M}_i \tag{3.36}$$

Substituting expressions for $\mathbf{d}_i$ and $\mathbf{d}_j$ and converting the cross-product to the matrix form, we get

$$\mathbf{J}_i\ddot{\boldsymbol{\alpha}}_i + \mathbf{R}_{pi}^\mathrm{T}(\mathbf{K}_p(\mathbf{x}_i + \mathbf{R}_{pi}\boldsymbol{\alpha}_i) - \mathbf{K}_p(\mathbf{x}_j + \mathbf{R}_{pj}\boldsymbol{\alpha}_j)) = \mathbf{M}_i \tag{3.37}$$

Expanding this, we get

$$\mathbf{J}_i\ddot{\boldsymbol{\alpha}}_i + \mathbf{R}_{pi}^\mathrm{T}\mathbf{K}_p\mathbf{x}_i + \mathbf{R}_{pi}^\mathrm{T}\mathbf{K}_p\mathbf{R}_{pi}\boldsymbol{\alpha}_i - \mathbf{R}_{pi}^\mathrm{T}\mathbf{K}_p\mathbf{x}_j - \mathbf{R}_{pi}^\mathrm{T}\mathbf{K}_p\mathbf{R}_{pj}\boldsymbol{\alpha}_j = \mathbf{M}_i \tag{3.38}$$

The moment equation may be written for body j as given in Equation 3.39, where $\mathbf{M}_j$ is the external moment acting on body j. Thus,

$$\mathbf{J}_j\ddot{\boldsymbol{\alpha}}_j - \mathbf{r}_j \times (\mathbf{K}_p\mathbf{d}_i - \mathbf{K}_p\mathbf{d}_j) = \mathbf{M}_j \tag{3.39}$$

Substituting $\mathbf{d}_i$ and $\mathbf{d}_j$ and converting the cross-product to the matrix form, we get

$$\mathbf{J}_j\ddot{\boldsymbol{\alpha}}_j - \mathbf{R}_{pj}^\mathrm{T}(\mathbf{K}_p(\mathbf{x}_i + \mathbf{R}_{pi}\boldsymbol{\alpha}_i) - \mathbf{K}_p(\mathbf{x}_j + \mathbf{R}_{pj}\boldsymbol{\alpha}_j)) = \mathbf{M}_j \tag{3.40}$$

Expanding this, we get

$$\mathbf{J}_j\ddot{\boldsymbol{\alpha}}_j - \mathbf{R}_{pj}^\mathrm{T}\mathbf{K}_p\mathbf{x}_i - \mathbf{R}_{pj}^\mathrm{T}\mathbf{K}_p\mathbf{R}_{pi}\boldsymbol{\alpha}_i + \mathbf{R}_{pj}^\mathrm{T}\mathbf{K}_p\mathbf{x}_j + \mathbf{R}_{pj}^\mathrm{T}\mathbf{K}_p\mathbf{R}_{pj}\boldsymbol{\alpha}_j = \mathbf{M}_j \tag{3.41}$$

If we then collect Equation 3.32 and Equation 3.38 for body I, then Equation 3.32 becomes

$$m_i\ddot{x}_i + \mathbf{K}_p\mathbf{x}_i + \mathbf{K}_p\mathbf{R}_{pi}\boldsymbol{\alpha}_i - \mathbf{K}_p\mathbf{x}_j - \mathbf{K}_p\mathbf{R}_{pj}\boldsymbol{\alpha}_j = \mathbf{F}_i \tag{3.42}$$

and Equation 3.38 becomes

$$\mathbf{J}_i\ddot{\boldsymbol{\alpha}}_i + \mathbf{R}_{pi}^\mathrm{T}\mathbf{K}_p\mathbf{x}_i + \mathbf{R}_{pi}^\mathrm{T}\mathbf{K}_p\mathbf{R}_{pi}\boldsymbol{\alpha}_i - \mathbf{R}_{pi}^\mathrm{T}\mathbf{K}_p\mathbf{x}_j - \mathbf{R}_{pi}^\mathrm{T}\mathbf{K}_p\mathbf{R}_{pj}\boldsymbol{\alpha}_j = \mathbf{M}_i \tag{3.43}$$

Expressing Equation 3.42 and Equation 3.43 in matrix form, we have

$$\begin{bmatrix} \mathbf{m}_i & 0 \\ 0 & \mathbf{J}_i \end{bmatrix} \begin{Bmatrix} \ddot{x}_i \\ \ddot{a}_i \end{Bmatrix} + \begin{bmatrix} \mathbf{K}_p & \mathbf{K}_p\mathbf{R}_{pi} \\ \mathbf{R}_{pi}^\mathrm{T}\mathbf{K}_p & \mathbf{R}_{pi}^\mathrm{T}\mathbf{K}_p\mathbf{R}_{pi} \end{bmatrix} \begin{Bmatrix} x_i \\ a_i \end{Bmatrix} - \begin{bmatrix} \mathbf{K}_p & \mathbf{K}_p\mathbf{R}_{pj} \\ \mathbf{R}_{pi}^\mathrm{T}\mathbf{K}_p & \mathbf{R}_{pi}^\mathrm{T}\mathbf{K}_p\mathbf{R}_{pj} \end{bmatrix} \begin{Bmatrix} x_j \\ a_j \end{Bmatrix} = \begin{Bmatrix} \mathbf{F}_i \\ \mathbf{M}_i \end{Bmatrix} \tag{3.44}$$

Similarly, if we collect Equation 3.34 and Equation 3.41 for body j, then Equation 3.34 becomes

$$m_j\ddot{x}_j - \mathbf{K}_p\mathbf{x}_i - \mathbf{K}_p\mathbf{R}_{pi}\boldsymbol{\alpha}_i + \mathbf{K}_p\mathbf{x}_j + \mathbf{K}_p\mathbf{R}_{pj}\boldsymbol{\alpha}_j = \mathbf{F}_j \tag{3.45}$$

and Equation 3.41 becomes

$$\mathbf{J}_j\ddot{\boldsymbol{\alpha}}_j - \mathbf{R}_{pj}^\mathrm{T}\mathbf{K}_p\mathbf{x}_i - \mathbf{R}_{pj}^\mathrm{T}\mathbf{K}_p\mathbf{R}_{pi}\boldsymbol{\alpha}_i + \mathbf{R}_{pj}^\mathrm{T}\mathbf{K}_p\mathbf{x}_j + \mathbf{R}_{pj}^\mathrm{T}\mathbf{K}_p\mathbf{R}_{pj}\boldsymbol{\alpha}_j = \mathbf{M}_j \tag{3.46}$$

Expressing Equation 3.45 and Equation 3.46 in the matrix form, we have

$$\begin{bmatrix} \mathbf{m}_j & 0 \\ 0 & \mathbf{J}_j \end{bmatrix} \begin{Bmatrix} \ddot{x}_j \\ \ddot{a}_j \end{Bmatrix} - \begin{bmatrix} \mathbf{K}_p & \mathbf{K}_p\mathbf{R}_{pi} \\ \mathbf{R}_{pj}^\mathrm{T}\mathbf{K}_p & \mathbf{R}_{pj}^\mathrm{T}\mathbf{K}_p\mathbf{R}_{pi} \end{bmatrix} \begin{Bmatrix} x_i \\ \alpha_i \end{Bmatrix} + \begin{bmatrix} \mathbf{K}_p & \mathbf{K}_p\mathbf{R}_{pj} \\ \mathbf{R}_{pj}^\mathrm{T}\mathbf{K}_p & \mathbf{R}_{pj}^\mathrm{T}\mathbf{K}_p\mathbf{R}_{pj} \end{bmatrix} \begin{Bmatrix} x_j \\ \alpha_j \end{Bmatrix} = \begin{Bmatrix} \mathbf{F}_j \\ \mathbf{M}_j \end{Bmatrix} \tag{3.47}$$

Overall, the equations of motion are now complete. Equation 3.44 and Equation 3.47 provide all that is needed to complete the final equations of motion. It is worth restating that the stiffness and damping matrices are identical in their structure. To obtain a damping matrix, all one needs to do is to replace the stiffness coefficients with the corresponding damping coefficients.

3.3 A Numerical Example

In order to illustrate the use of the equations given before, let us consider a rigid body flexibly supported by a number of springs. For this, the simplest starting point would be Equation 3.44

$$\begin{bmatrix} \mathbf{m}_i & 0 \\ 0 & \mathbf{J}_i \end{bmatrix} \begin{Bmatrix} \ddot{x}_i \\ \ddot{a}_i \end{Bmatrix} + \begin{bmatrix} \mathbf{K}_p & \mathbf{K}_p\mathbf{R}_{pi} \\ \mathbf{R}_{pi}^\mathrm{T}\mathbf{K}_p & \mathbf{R}_{pi}^\mathrm{T}\mathbf{K}_p\mathbf{R}_{pi} \end{bmatrix} \begin{Bmatrix} x_i \\ a_i \end{Bmatrix} - \begin{bmatrix} \mathbf{K}_p & \mathbf{K}_p\mathbf{R}_{pj} \\ \mathbf{R}_{pi}^\mathrm{T}\mathbf{K}_p & \mathbf{R}_{pi}^\mathrm{T}\mathbf{K}_p\mathbf{R}_{pj} \end{bmatrix} \begin{Bmatrix} x_j \\ a_j \end{Bmatrix} = \begin{Bmatrix} \mathbf{F}_i \\ \mathbf{M}_i \end{Bmatrix} \tag{3.48}$$

Since body j does not exist, all the terms relevant to body j will disappear. Furthermore, since we are dealing with a single mass, the suffix i is not needed either. However, for n number of springs, the stiffness matrices need to be summed-up. Summation has to be carried out for each stiffness p attached at a position on the body. We then have

$$\begin{bmatrix} \mathbf{m} & \mathbf{0} \\ \mathbf{0} & \mathbf{J} \end{bmatrix} \begin{Bmatrix} \ddot{x} \\ \ddot{\alpha} \end{Bmatrix} + \begin{bmatrix} \displaystyle\sum_{p=1}^{n} \mathbf{K}_p & \displaystyle\sum_{p=1}^{n} \mathbf{K}_p \mathbf{R}_p \\ \displaystyle\sum_{p=1}^{n} \mathbf{R}_p^{\mathrm{T}} \mathbf{K}_p & \displaystyle\sum_{p=1}^{n} \mathbf{R}_p^{\mathrm{T}} \mathbf{K}_p \mathbf{R}_p \end{bmatrix} \begin{Bmatrix} x \\ \alpha \end{Bmatrix} = \begin{Bmatrix} \mathbf{F} \\ \mathbf{M} \end{Bmatrix} \tag{3.49}$$

For a situation where the axes of the springs are parallel to the global coordinate system, no transformation of the stiffness matrix is needed. Hence, $\mathbf{k}_p = \mathbf{K}_p$. To obtain the submatrices of the stiffness matrix given in Equation 3.49, start with the stiffness matrix for spring p. Specifically,

$$\mathbf{K}_p = \begin{bmatrix} k_{px} & 0 & 0 \\ 0 & k_{py} & 0 \\ 0 & 0 & k_{pz} \end{bmatrix} \tag{3.50}$$

Now, $\mathbf{K}_p \mathbf{R}_p$ is given by

$$\mathbf{K}_p \mathbf{R}_p = \begin{bmatrix} k_{px} & 0 & 0 \\ 0 & k_{py} & 0 \\ 0 & 0 & k_{pz} \end{bmatrix} \begin{bmatrix} 0 & z_p & -y_p \\ -z_p & 0 & x_p \\ y_p & -x_p & 0 \end{bmatrix} \tag{3.51}$$

Expanding this, we get

$$\mathbf{K}_p \mathbf{R}_p = \begin{bmatrix} 0 & k_{px} z_p & -k_{px} y_p \\ -k_{py} z_p & 0 & k_{py} x_p \\ k_{pz} y_p & -k_{pz} x_p & 0 \end{bmatrix} \tag{3.52}$$

For $\mathbf{R}_p^{\mathrm{T}} \mathbf{K}_p$, we have

$$\mathbf{R}_p^{\mathrm{T}} \mathbf{K}_p = \begin{bmatrix} 0 & -z_p & y_p \\ z_p & 0 & -x_p \\ -y_p & x_p & 0 \end{bmatrix} \begin{bmatrix} k_{px} & 0 & 0 \\ 0 & k_{py} & 0 \\ 0 & 0 & k_{pz} \end{bmatrix} \tag{3.53}$$

Expanding this, we get

$$\mathbf{R}_p^{\mathrm{T}} \mathbf{K}_p = \begin{bmatrix} 0 & -k_{py} z_p & k_{pz} y_p \\ k_{px} z_p & 0 & -k_{pz} x_p \\ -k_{px} y_p & k_{py} x_p & 0 \end{bmatrix} \tag{3.54}$$

Finally, $\mathbf{R}_p^{\mathrm{T}} \mathbf{K}_p \mathbf{R}_p$ is given by

$$\mathbf{R}_p^{\mathrm{T}} \mathbf{K}_p \mathbf{R}_p = \begin{bmatrix} 0 & -k_{py} z_p & k_{pz} y_p \\ k_{px} z_p & 0 & -k_{pz} x_p \\ -k_{px} y_p & k_{py} x_p & 0 \end{bmatrix} \begin{bmatrix} 0 & z_p & -y_p \\ -z_p & 0 & x_p \\ y_p & -x_p & 0 \end{bmatrix} \tag{3.55}$$

$$\mathbf{R}_p^{\mathrm{T}}\mathbf{K}_p\mathbf{R}_p = \begin{bmatrix} k_{pz}y_p^2 + k_{py}z_p^2 & -k_{pz}x_py_p & -k_yx_pz_p \\ -k_{pz}x_py_p & k_{pz}x_p^2 + k_{px}z_p^2 & -k_{px}y_pz_p \\ -k_{py}x_pz_p & -k_{px}y_pz_p & k_{py}x_p^2 + k_{px}y_p^2 \end{bmatrix} \qquad (3.56)$$

The overall stiffness matrix from Equation (8.48) for a single spring is given by

$$\begin{bmatrix} k_{px} & 0 & 0 & 0 & k_{px}z_p & -k_{px}y_p \\ 0 & k_{py} & 0 & -k_{py}z_p & 0 & k_{py}x_p \\ 0 & 0 & k_{pz} & k_{pz}y_p & -k_{pz}x_p & 0 \\ 0 & -k_{py}z_p & k_{pz}y_p & k_{pz}y_p^2 + k_{py}z_p^2 & -k_{pz}x_py_p & -k_{py}x_pz_p \\ k_{px}z_p & 0 & -k_{pz}x_p & -k_{pz}x_py_p & k_{pz}x_p^2 + k_{px}z_p^2 & -k_{px}y_pz_p \\ -k_{px}y_p & k_{py}x_p & 0 & -k_{py}x_pz_p & -k_{px}y_pz_p & k_{py}x_p^2 + k_{px}y_p^2 \end{bmatrix} \qquad (3.57)$$

The mass matrix is diagonal and, for the inertia matrix, it is assumed that the principal axes of the body coincide with the global coordinate system. Specifically,

$$\mathbf{m} = \begin{bmatrix} m & 0 & 0 \\ 0 & m & 0 \\ 0 & 0 & m \end{bmatrix} \qquad (3.58)$$

$$\mathbf{J} = \begin{bmatrix} Ixx & 0 & 0 \\ 0 & Iyy & 0 \\ 0 & 0 & Izz \end{bmatrix} \qquad (3.59)$$

Now, the overall equations of motion may be assembled for n springs.

$$\begin{bmatrix} m & 0 & 0 & 0 & 0 & 0 \\ 0 & m & 0 & 0 & 0 & 0 \\ 0 & 0 & m & 0 & 0 & 0 \\ 0 & 0 & 0 & Ixx & 0 & 0 \\ 0 & 0 & 0 & 0 & Iyy & 0 \\ 0 & 0 & 0 & 0 & 0 & Izz \end{bmatrix} \begin{Bmatrix} \ddot{x} \\ \ddot{y} \\ \ddot{z} \\ \ddot{\alpha} \\ \ddot{\beta} \\ \ddot{\gamma} \end{Bmatrix} +$$

$$\sum_{p=1}^{n} \begin{bmatrix} k_{px} & 0 & 0 & 0 & k_{px}z_p & -k_{px}y_p \\ 0 & k_{py} & 0 & -k_{py}z_p & 0 & k_{py}x_p \\ 0 & 0 & k_{pz} & k_{pz}y_p & -k_{pz}x_p & 0 \\ 0 & -k_{py}z_p & k_{pz}y_p & k_{pz}y_p^2 + k_{py}z_p^2 & -k_{pz}x_py_p & -k_{py}x_pz_p \\ k_{px}z_p & 0 & -k_{pz}x_p & -k_{pz}x_py_p & k_{pz}x_p^2 + k_{px}z_p^2 & -k_{px}y_pz_p \\ -k_{px}y_p & k_{py}x_p & 0 & -k_{py}x_pz_p & -k_{px}y_pz_p & k_{py}x_p^2 + k_{px}y_p^2 \end{bmatrix} \begin{Bmatrix} x \\ y \\ z \\ \alpha \\ \beta \\ \gamma \end{Bmatrix} = \begin{Bmatrix} F_x \\ F_y \\ F_z \\ M_x \\ M_y \\ M_z \end{Bmatrix}$$

$$(3.60)$$

3.3.1 A Uniform Rectangular Prism

A rectangular prism is supported by four springs as shown in Figure 3.3. Springs have stiffness values in all three directions (k_{px}, k_{py}, k_{pz}, where p is the spring number). The axes of each spring in which the stiffness values are measured are parallel to the principal axes of the springs, which in turn are parallel to the global coordinate system of the rectangular prism. Thus, no transformation is needed. The end of spring p is located at (x_p, y_p, z_p), measured relative to the COG of the body. The mass of the prism is m and the principal moments of inertia are Ixx, Iyy, and Izz. A simplified equation of motion of the system in 3-D space may be obtained from Equation 3.59. If one attempts to

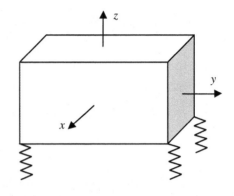

FIGURE 3.3   A rectangular prism supported on springs.

carry this out, one will realize that some terms will disappear because the z components of the positions are all zero and some will disappear because of the symmetry of points.

The body shown in Figure 3.3 corresponds to $m = 1000$ kg, moments of inertia $Ixx = 10$ kg m², $Iyy = 20$ kg m², and $Izz = 30$ kg m², supported by four identical (thus, point suffix p is dropped) springs with stiffness values ($kx = 10{,}000$ N/m, $ky = 20{,}000$ N/m, $kz = 30{,}000$ N/m). The positions of the springs are given as follows:

$$P1(1, 2, 0)$$

$$P2(1, -2, 0)$$

$$P3(-1, 2, 0)$$

$$P4(-1, -2, 0)$$

The coordinates imply that the COG is on the bottom plane of the prism. The system has six degrees of freedom and all six natural frequencies will be calculated.

Since stiffness parameters are on the Oxy plane, no coupling will occur between (x and β) and (y and α). Similarly, the vertical motion is also uncoupled from the others due to symmetry. Thus,

$$\omega_x = \sqrt{\frac{\sum_{p=1}^{4} k_{px}}{m}} = \sqrt{\frac{40{,}000}{1000}} = 6.32 \text{ rad/sec} = 1.0065 \text{ Hz}$$

$$\omega_y = \sqrt{\frac{\sum_{p=1}^{4} k_{py}}{m}} = \sqrt{\frac{80{,}000}{1000}} = 8.94 \text{ rad/sec} = 1.4235 \text{ Hz}$$

$$\omega_z = \sqrt{\frac{\sum_{p=1}^{4} k_{pz}}{m}} = \sqrt{\frac{120{,}000}{1000}} = 10.95 \text{ rad/sec} = 1.743 \text{ Hz}$$

$$\omega_\alpha = \sqrt{\frac{\sum_{p=1}^{4} y_p^2 k_{pz}}{I_{xx}}} = \sqrt{\frac{4 \times 2^2 \times 30{,}000}{10}} = 219.09 \text{ rad/sec} = 34.87 \text{ Hz}$$

$$\omega_\beta = \sqrt{\frac{\sum_{p=1}^{4} x_p^2 k_{pz}}{I_{yy}}} = \sqrt{\frac{4 \times 1^2 \times 30,000}{20}} = 77.45 \text{ rad/sec} = 12.32 \text{ Hz}$$

$$\omega_\gamma = \sqrt{\frac{\sum_{p=1}^{4} (x_p^2 k_{py} + y_p^2 k_{px})}{I_{zz}}} = \sqrt{\frac{4(1^2 \times 20,000 + 2^2 \times 10,000)}{30}} = 89.44 \text{ rad/sec} = 14.23 \text{ Hz}$$

3.3.2   VIBRATIO Output

For the numerical problem given above, the output obtained from the software package VIBRATIO (see Appendix 3A for the full listing) is tabulated below.

Dominant Oscillation Direction	Frequencies
Frequency in X	1.01 Hz (60 CPM)
Frequency in Y	1.42 Hz (85 CPM)
Frequency in Z	1.74 Hz (105 CPM)
Frequency in alpha	34.87 Hz (2092 CPM)
Frequency in beta	12.33 Hz (740 CPM)
Frequency in gamma	14.24 Hz (854 CPM)

3.4   An Industrial Vibration Design Problem

In the analysis of vibration characteristics of an engine or an engine-generator set, flexibly supported rigid-body representations are commonly used. In the case study given here, an engine assembly system is considered. In designing an engine mounting system, engineers consider a number of issues. Particularly useful are: static deflection; natural frequencies and spread of these frequencies relative to the engine speed range; time response under shock or other transient loads; and frequency response of the system, especially when subjected to unbalanced engine-generated excitations.

3.4.1   Static Deflection

This problem involves the selection of mounting geometry and mount stiffness parameters. Here, the design engineer seeks to achieve a pure deflection with as little tilt as possible. In principle, there is nothing wrong with some mounts deflecting more than the others, but an excessive tilt normally indicates other problems such as a high degree of coupling between the modes of motion. However, achieving a pure vertical deflection, at least mathematically, is relatively easy and involves the selection of spring positions/stiffness values to satisfy the following conditions:

$$\sum_{p=1}^{n\text{-spring}} y_p k_{pz} = 0$$

and

$$\sum_{p=1}^{n\text{-spring}} x_p k_{pz} = 0.$$

Here, x_p and y_p are spring mount positions and k_{pz} is the spring stiffness in the z direction. In addition, the designer needs to ensure that the static deflection is well within the allowable deflection range, especially since it needs to account for any additional deflection due to vibration.

3.4.2 Natural Frequencies

Normally, design application demands that the natural frequencies are kept away from the running speed of an engine. Such a requirement is easy to satisfy for single-DoF systems. However, in real-life situations, the number of DoF(s) is six just for a single rigid body. Any change to the mounting configuration or stiffness parameters (or mass distribution) will affect all six natural frequencies. In order to be able to modify a natural frequency corresponding to a particular mode shape (e.g., resonance in the vertical direction), a vibration design engineer will attempt to decouple various modes of oscillation. Such a design requirement, however, is unrealistic for many engineering problems, even for a single rigid body, due to space considerations and is practically impossible for multiple rigid body systems. However, it is possible to achieve partial decoupling. For example, the condition given above for pure vertical deflection will also provide decoupling between the rocking motion about two-coordinate systems (about Ox and Oy) in the horizontal plane and vertical motion.

3.4.3 Transient Response Analysis

Static deflection analysis and modal analysis are normally essential. However, there are two more problem-specific analyses that a design engineer may need to perform depending on the problem. For example, in many engineering applications, the response of a flexibly supported system to a shock loading is very important. In this case, coupling among modes as well as stiffness of the system play an important role in determining the levels of shock transmission to the engine system. As a rule of thumb, the softer the spring the smaller the shock transferred to the flexibly mounted structure. Making mounting stiffness elements softer may not be a realistic option as this can result in unacceptable static deflections and may even be in conflict with the requirements discussed above in relation to the positioning of a natural frequency relative to the operating speed of the engine and the static deflection.

3.4.4 Frequency Analysis

Frequency analysis is probably one of the most useful among the various types of analysis listed above. If the vibration problem is not transient, then it is a problem involving a steady-state vibration. The analysis of a steady-state vibration problem under sinusoidal excitation is normally referred to as frequency analysis. In mathematical terms, this is the particular solution of the differential equation of motion. Under a sinusoidal excitation, a vibrating system reaches a steady-state vibration and frequency analysis provides information of the amplitude of vibration as a function of excitation frequency. If there is more than one excitation force, then the resulting motion will be a combination of the results due to each excitation. Multiexcitation force analysis is sometimes referred to as harmonics analysis. Here, the designer needs to ensure that the highest amplitude of oscillation under harmonic excitation does not exceed the safe limits for the system and, in particular, for the mounts.

Although design considerations linked to static deflection, positioning of natural frequencies, decoupling of modes, and response to shock represent a large proportion of vibration design problems, there are other and more complex design specifications. For example, minimizing vibration at a point on a flexibly supported body may be considered a design objective. Such a requirement may then cause problems where the engineer intends to place a drive shaft coupling or an additional mounting at this position. Because the vibration (or deflection) is at its minimum at the assembly point, a coupling or additional mount will add a minimum constraint to the system.

Obviously, all the design objectives discussed above have to be satisfied within the physical constraints relating to the problem at hand. In general, these constraints are associated with space and the stiffness range of industrially available mountings.

3.4.5 A Flexibly Supported Engine — A Numerical Problem

The example considered here involves a study of an engine-mounting configuration. Mounting configurations are restricted by the geometry and little flexibility exists in modifying these positions. They are located by the engine manufacturer. To perform vibration analysis, the mass/moments of inertia values of the system, the coordinates of the mounting positions relative to the COG, and the forces acting on the system are needed. Although the mass is normally given or easy to obtain, moments of inertia and the COG are not always supplied by engine manufacturers. Even though it is possible to build a solid model of an engine in order to calculate moments of inertia, this is a rather tedious and costly task. The alternative is to define engine moments of inertia in an approximate manner. This can be done by assuming that the main assemblies of the engine are made of regular geometrical primitives representing approximate shapes without going into exact geometrical models. Such an approach works reasonably well, especially since the mass values of these primitives can be obtained in an exact manner. When calculating the moments of inertia and the overall mass of the assembly, its COG can also be calculated. Once these are obtained, the mount positions can be calculated relative to the COG. Having obtained the mass, the moments of inertia, the COG, and the mount position coordinates relative to the COG, the main step of analysis may be started. This involves selecting the mount stiffness parameters in such a way that the various conditions and objectives described above are met. The vibration design, like all engineering problems, involves reconciling many conflicting requirements.

Engine mass and moments of inertia (symmetry of mass distribution is assumed)
$m = 250$ kg and moments of inertia $Ixx = 45$ kg m^2, $Iyy = 80$ kg m^2, and $Izz = 110$ kg m^2.

Mounts stiffness values:
  ($kx = 150{,}000$ N/m, $ky = 150{,}000$ N/m, $kz = 300{,}000$ N/m).

Mount positions (all in mm):

1	0.000	100.000	− 75.000
2	250.000	0.000	− 75.000
3	0.000	− 100.000	− 75.000
4	− 170.000	0.000	− 75.000

3.4.5.1 Satisfying Static Deflection

On running a static analysis under a vertical load of 2500 N (weight), the following results are obtained:
  (displacements are in mm and angles are in rad)

X	Y	Z	Alpha	Beta	Gamma	xc	yc	zc
0.1392	0.0000	2.1205	0.0000	0.0019	0.0000	1142.5000	0.0000	− 75.0000

In relation to the static deflection, there are two considerations: (i) overall deflection should not be more than what is allowed by the deflection range of the springs selected for the design, and (ii) the static position and orientation of the engine should not be outside what is allowed by spatial and other constraints; i.e., it should not tilt to one side excessively. In either case, stiffer springs will tend to solve the problem. However, such a choice may not be the best for transients, shocks, and vibration transmission from the supporting frame. "Tilt" level calculated above is assumed to be small (0.0019 rad).

The results from the program also list the deflection at the mount positions (in mm) as: mass no. = 1

Mount no.	Position			Deflection		
	X	Y	Z	x	y	z
1	0.000	100.000	−75.000	0.000	0.000	2.120
2	250.000	0.000	−75.000	0.000	0.000	1.656
3	0.000	−100.000	−75.000	0.000	0.000	2.120
4	−170.000	0.000	−75.000	0.000	0.000	2.436

The maximum deflection is at the fourth mount position and is 2.436 mm. The mount selected should give a deflection range that would extend well beyond this to ensure that any additional deflection due to vibration could be accommodated.

The coordinates xc, yc, zc give the instantaneous center of rotation for this particular static deflection result. This point (as discussed above) may be used in some design applications as it remains stationary during the deflection.

3.4.5.2 Eigenvalue Analysis

The eigenvalue analysis will help ensure that the natural frequencies are not in the vicinity of the idling speed of the engine. It may also help to minimize the number of natural frequencies in the speed range of the engine.

Since the spring positions and their locations are already specified to satisfy the considerations for static deflection, it becomes difficult to modify them to satisfy the "natural frequency" requirements as well. However, all stiffness values could be increased together in the same proportion. This ensures that the "no tilt" condition is maintained. Of course, stiffening now reduces the static deflection and is likely to increase the vibration transmission to the frame.

The eigenvalue analysis results are listed below. Here, the natural frequencies spread from 1.98 to 12.09 Hz. The widest gap between these frequencies is between 3.17 and 8.69 Hz. It would be desirable to have the idling speed in the middle of this range. It is equally important that the cruising speed does not coincide with the two higher frequencies.

X	Y	Z	Alpha	Beta	Gamma
		Frequency in X = 8.63 Hz (518 CPM)			
1.0000	0.0000	−0.0111	0.0000	−0.2709	0.0000
		Frequency in Y = 8.69 Hz (521 CPM)			
0.0000	1.0000	0.0000	0.4403	0.0000	0.0484
		Frequency in Z = 12.09 Hz (725 CPM)			
−0.0051	0.0000	−1.0000	0.0000	0.0685	0.0000
		Frequency in alpha = 1.98 Hz (119 CPM)			
0.0000	0.0797	0.0000	−1.0000	0.0000	−0.0226
		Frequency in beta = 3.17 Hz (190 CPM)			
0.0869	0.0000	0.0215	0.0000	1.0000	0.0000
		Frequency in gamma = 2.13 Hz (128 CPM)			
0.0000	−0.0163	0.0000	−0.0625	0.0000	1.0000

In order to achieve decoupling between different motions, one practical technique is to minimize the distance between the COG and the "center of stiffness." Center of stiffness is a crude term used in industry to ensure that the coupling between different motions of body is minimized. The definition of center of stiffness is similar to that of the COG. The Ox axis is located in such a way that $\sum_p y_p k_{pz} = 0$ holds. Then this axis will pass through the center of stiffness. Similarly, $\sum_p x_p k_{pz} = 0$ will hold for the Oy axis passing through the center of stiffness. If the center of stiffness coincides with the COG and the axes defined by the first moment of stiffness coincide with the principal axis of mass, then full decoupling can be achieved. As far as the horizontal plane is concerned, this also ensures that the assembly is leveled. Note that the relationships are the same as the "no tilt" condition described above for static deflection. Normally, it may not be possible to achieve this in all three planes and the designer may choose to achieve this in one plane where the excitation forces are the greatest. However, it is common among designers to focus on the horizontal plane alone, purely due to deflection under gravity considerations.

3.4.5.3 Time Domain Analysis — Analysis of the System under a Shock Loading

Suppose that a 10 g, 10 msec, half sine shock is applied in the vertical, z, direction. The shock response in the z direction is shown in Figure 3.4. The shock response in the y direction is shown in Figure 3.5.

The results show that, in addition to static deflection, if the mounts were to withstand the applied shock, they should be able to deflect 9 mm in shear and more than 6 mm in the vertical directions.

3.4.5.4 Frequency Analysis

The frequency analysis specifications are given below.

The engine is subjected to an unbalanced force which is known to be proportional to the square of the engine running speed. In other words, this is given as Aw^2. It is measured that when the engine speed is 300 rpm the unbalanced force is 250 N. A simple calculation shows that $A = 1.013$. The force menu option 20 in VIBRATIO provides the required excitation, which increases with the square of the running speed. The option 20 allows the A value to be linearly increased between the start and end frequencies during which the excitation is active. In our case, this is to be taken to be the same (frequency is independent of the A value). The vertical amplitude vs. frequency results are given in Figure 3.6. The amplitudes in other directions are much smaller and are not shown here. The analysis is not carried out beyond 15 Hz as we know already that the maximum resonance is at 12.09 Hz. According to the result, the selected mount should allow a 16 mm deflection on top of static deflection. Now, the designer should be in a position to make a decision on whether the selected

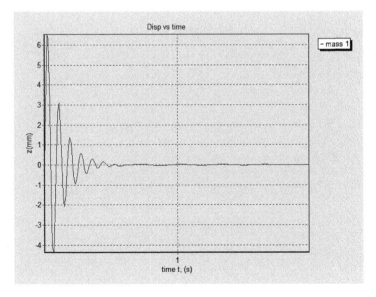

FIGURE 3.4 Shock response in the z direction.

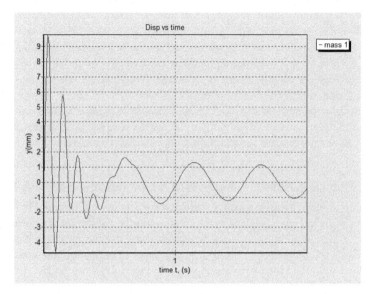

FIGURE 3.5   Shock response in the y direction.

spring type is acceptable or not. If the answer is no, then the whole analysis process has to be repeated to find an acceptable solution.

3.5   Programming Considerations

Equations developed in this chapter are formulated and structured in such a way that they can be used for developing general vibration analysis software. As the equations given above refer to bodies i and j, they can be placed in the global coordinates accordingly. Four submatrices (6×6) will be placed as follows:

$$\begin{bmatrix} \mathbf{K}_r & \mathbf{K}_r \mathbf{R}_{pi} \\ \mathbf{R}_{pi}^{\mathrm{T}} \mathbf{K}_r & \mathbf{R}_{pi}^{\mathrm{T}} \mathbf{K}_r \mathbf{R}_{pi} \end{bmatrix} \text{ will be placed, starting from position } (6(i-1)+1, 6(i-1)+1)$$

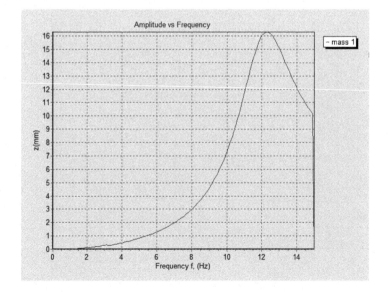

FIGURE 3.6   Frequency response in the z direction for unbalanced engine excitation.

$$-\begin{bmatrix} \mathbf{K}_r & \mathbf{K}_r\mathbf{R}_{pj} \\ \mathbf{R}_{pi}^{\mathrm{T}}\mathbf{K}_r & \mathbf{R}_{pi}^{\mathrm{T}}\mathbf{K}_r\mathbf{R}_{pj} \end{bmatrix}$$ will be placed, starting from position $(6(i-1)+1, 6(j-1)+1)$

$$-\begin{bmatrix} \mathbf{K}_r & \mathbf{K}_r\mathbf{R}_{pi} \\ \mathbf{R}_{pj}^{\mathrm{T}}\mathbf{K}_r & \mathbf{R}_{pj}^{\mathrm{T}}\mathbf{K}_r\mathbf{R}_{pi} \end{bmatrix}$$ will be placed, starting from position $(6(j-1)+1, 6(i-1)+1)$

$$\begin{bmatrix} \mathbf{K}_r & \mathbf{K}_r\mathbf{R}_{pj} \\ \mathbf{R}_{pj}^{\mathrm{T}}\mathbf{K}_r & \mathbf{R}_{pj}^{\mathrm{T}}\mathbf{K}_r\mathbf{R}_{pj} \end{bmatrix}$$ will be placed, starting from position $(6(j-1)+1, 6(j-1)+1)$

Entry to the global stiffness matrix must be additive (a new entry is added to the previously entered values to account for any contribution coming from other springs). In programming terms, the global stiffness matrix is constructed based on the spring count loop. Therefore, the data structure for a spring must be such that the end of the spring refers to the mass number. Hence, for a spring stiffness, submatrices could be located in the global matrix according to the mass numbers to which the ends of that spring are attached.

To start assembling the equations, the first step is to express the stiffness matrices in global coordinates. Here, it is assumed that the stiffness values are given in three orthogonal axes. These axes will be assumed to be the principal axes of the springs (in practice, this is not strictly correct and the assumption holds only for small deflections). The orientation of the principal axes relative to the global coordinates can be described in terms of Euler angles (or using direction cosines). Once the transformation matrix is obtained, then the stiffness can be expressed in the global coordinate system Equation 3.4. In developing one's own software, it is important to bear in mind the way the transformation matrix is constructed. Euler angles have to be applied one by one and the order in which they are processed is important. For example, to orient a given body, one may assume that, initially, its local coordinates are parallel to the principal coordinate frame. From here, the first Euler angle will rotate the body to a new orientation. Relative to this position, the next rotation will be applied, resulting in yet another orientation. Lastly, the third rotation will be applied relative to this orientation, giving the final position of the body under consideration. Assuming that these rotations are α β γ in the given order, the transformation matrix is built as

$$T = T(\alpha)T(\beta)T(\gamma)$$

In other words, any vector in the local coordinates will be transformed to the global coordinates in reverse order. To explain further, any vector in the local coordinates is first transformed by $T(\gamma)$, then the resulting vector is transformed by $T(\beta)$, and finally the resulting vector is transformed by $T(\alpha)$. The expression $\mathbf{X} = \mathbf{Tx}$ processes these matrices in that order.

Transformation will apply to spring stiffness matrices and inertia matrices. Therefore, the input to the vibration program has to account for Euler angles for each body and spring, and the order in which these rotations are performed.

Once the transformation is completed, all that remains is to perform the appropriate matrix algebra in order to construct the stiffness matrix for each spring and its location in the global stiffness matrix. The inertia matrix for each mass, after expressing it in the global coordinate frame by performing matrix transformation, will need to be placed in the appropriate location in the global inertia matrix.

3.6 VIBRATIO

3.6.1 Capabilities

VIBRATIO is a vibration analysis program designed to model flexibly supported multibody systems. An educational version is made available free of charge by the author of this chapter for the readers of this handbook. The executable file could be downloaded from the website, www.signal-research.com. The educational version will allow analysis of a flexibly supported system not exceeding more than 10 masses (60 dof). For the purpose of supporting the theory presented in this handbook, four types of analysis are activated: static; eigenvalue (both damped and undamped); frequency domain; and time domain. Standard excitations for time and frequency domains will be based on the excitation functions available in the menu. Numerically defined excitations, and all the other modules, such as signal processing, finite-element shaft analysis, fatigue analysis, mass design, etc. are not available in this version.

3.6.2 Modeling on VIBRATIO

Click vibratio2002 from start/programs/vibratio2002 menu. The VIBRATIO screen shown in Figure 3.7 will be displayed. The top part of the frame gives inputs for application details. The date/time box can be double-clicked to produce current date and time. File options are listed in the file box. Only files with a VIB extension are filtered for the listing. You may also use the file menu to load VIBRATIO files. If you use any other extension to save your data, you will have to use the file menu to open it. VIB files are text files and can be read by any text editor. It is not advisable, however, to create these manually by using any editor other than VIBRATIO.

3.6.2.1 Entering Spring Data

The second section on this screen, titled "Spring Type Description," is for entering spring stiffnesses. Here, the *types* of springs are defined, not the definition of individual springs.

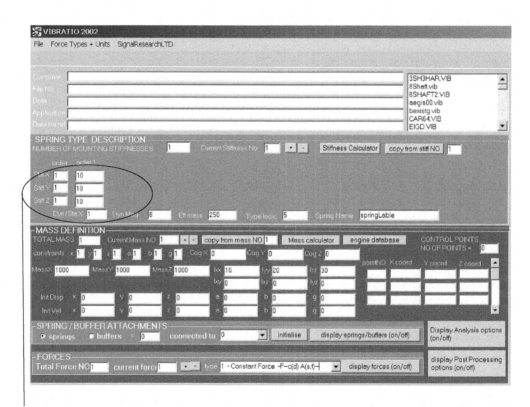

Stiffnesses kx, ky and kz.

FIGURE 3.7 VIBRATIO main window.

Order should always be set to 1. In general, it is assumed that the spring deflections can be described as polynomials, and "order" refers to order of the polynomial. Only linear capability is offered in the present version; thus, order must always be set to 1.

3.6.2.2 Entering Mass Data

When VIBRATIO runs, it automatically loads default mass data. To create a two-mass system, change "total mass" to 2. This is shown in Figure 3.8.

Normally, if these two masses are not the same then you will need to modify accordingly. Also, you may copy mass data from one previously defined mass to another by clicking the "copy from" button. CogX, CogY, CogZ are the centers of mass of the current mass relative to the COG of mass 1. It means that the program will not allow you to change the COG coordinates of mass 1. "Frame" shows that you have mass$X$, mass$Y$, mass$Z$ values. For normal applications, mass does not have a directional property and values in each direction will be the same. There are applications where directional "effective mass" may be appropriate, such as modeling the ground mass for earthquake analysis.

Moment of inertia elements are calculated relative to the COG of current mass and about coordinates that are parallel to the coordinates of the global frame. The center of the global coordinate system coincides with the center of mass 1. If the principal coordinate system of the current body is not parallel to the global coordinate system, then cross-inertias Ixy, Ixz, Iyz need to be calculated. The full version of

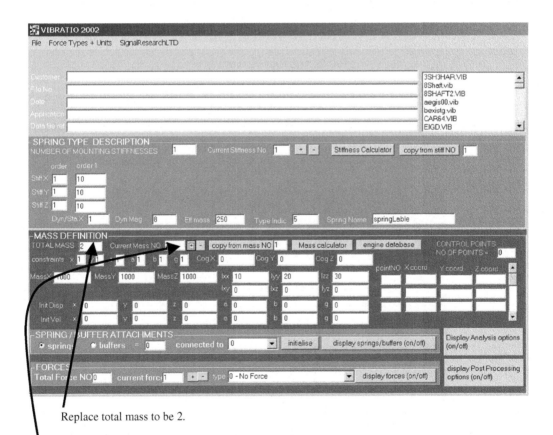

Replace total mass to be 2.

Click "+" button to view mass 2 properties.

FIGURE 3.8 Mass data entry.

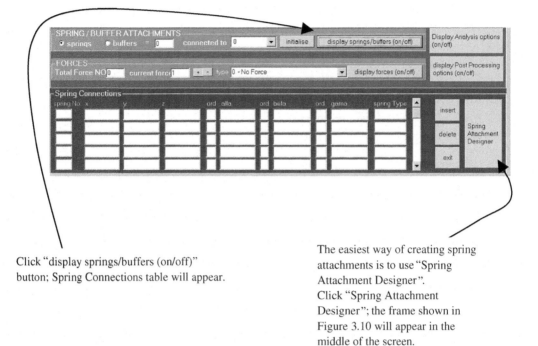

Click "display springs/buffers (on/off)"
button; Spring Connections table will appear.

The easiest way of creating spring
attachments is to use "Spring
Attachment Designer".
Click "Spring Attachment
Designer"; the frame shown in
Figure 3.10 will appear in the
middle of the screen.

FIGURE 3.9   Entering spring data.

VIBRATIO offers a mass calculator that, among other things, can calculate the necessary transformation
for given sets of Euler angles.

Control points: Here, you may enter the coordinates of points at which the linear deflection module will
calculate the displacements. For time and frequency domain analyses this module is not used. Graphics
programs come with options to create point tables for analyzing motions at those points.

FIGURE 3.10   Spring attachment designer window.

Initial conditions: For each mass, initial conditions may be assigned (displacements and velocities) for time-domain analysis.

Now you have a two-mass system where both masses are identical.

3.6.2.3 Entering Spring Attachment Data

The process of entering spring data is illustrated in Figure 3.9 to Figure 3.11. The first step is to decide how these bodies are connected to the ground and to each other. Connection can be by a spring or a buffer. Buffers are not covered in this chapter. One needs to make sure that the spring option button is selected (spring option is selected as the default). Consider the option buttons. The "text box" refers to the total number of connections to be entered in this frame. Do not modify this one for now. What you enter in this box will depend on how you create these connections. The next very important action is to select "connected to" combo box. This refers to the mass number to which the current mass is connected. If the connection is to the ground, then "0" should be selected. Ground is called mass 0. Once you select a connection mass, "+" will appear in front of the mass number. This makes it easy to identify the masses to which a connection is made. If a "+" character had not been added, then the user would have to search all the mass numbers one by one to view the data. Even in a 10-mass system, this could involve checking 55 possible connections. Note that "connected to" lists mass numbers that are less than the current mass. For example, if the current mass number is 5, then the "connected to" list will have 4, 3, 2, 1, 0.

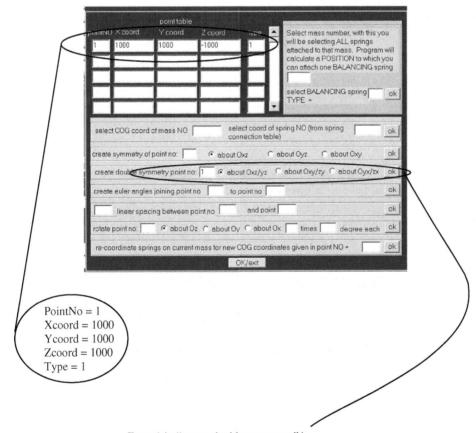

PointNo = 1
Xcoord = 1000
Ycoord = 1000
Zcoord = 1000
Type = 1

Enter 1 in "create double symmetry" in
sub-frame and click "about Oxz/yz" option.

FIGURE 3.11 Spring data entry.

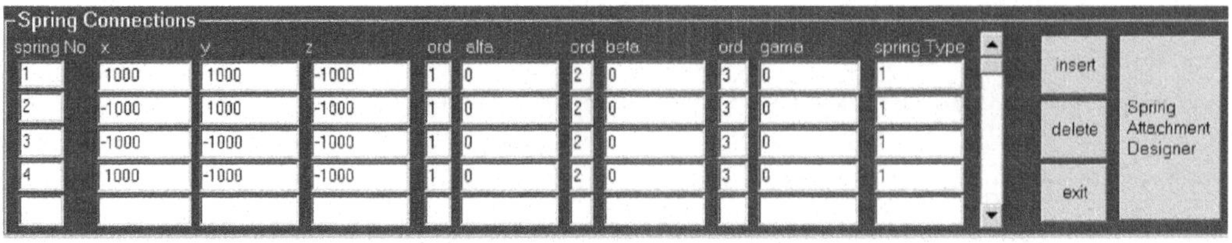

spring No	x	y	z	ord	alfa	ord	beta	ord	gama	spring Type
1	1000	1000	-1000	1	0	2	0	3	0	1
2	-1000	1000	-1000	1	0	2	0	3	0	1
3	-1000	-1000	-1000	1	0	2	0	3	0	1
4	1000	-1000	-1000	1	0	2	0	3	0	1

Spring Connections

insert

delete

Spring Attachment Designer

exit

FIGURE 3.12   Four spring attachments generated by "spring attachment designer."

To start entering connections, make sure that the current mass is 1, the "spring" rather than the "buffer" option is selected, and, in this case, "connected to" would be 0 (the only option). Also, make sure that you have pressed the "display springs/buffers" (on/off) button. Now you have three options:

1. Enter the total number of spring connections and start entering the data one by one, but make sure that you enter no more than the total connections you had specified. (This is rather tedious and most of the spring attachments will have identical angle values.)
2. Once you have selected the total number of springs, click the "initialize" button. This will fill all the cells with default numbers and you can simply modify these to create your own data.
3. An even easier approach is to follow the steps given below.

Enter values as shown in Figure 3.11. Point number is used as a reference in this frame only. Coordinates X, Y, Z are in mm and are relative to the body displayed currently and connecting to the "connected to" mass. If you want to attach springs to the next mass, you must display the mass in your main window (attachment is to the current mass). "Type" is the spring type. Spring type numbers were previously described.

Double symmetry means the point (point number 1, referring to the point above) will be duplicated by its symmetry about the Oxz plane. Then these two points will be doubled (four altogether) by taking their symmetry about Oyz. Click "ok" (for double symmetry) to complete the operation. The points shown in Figure 3.12 are now generated. Note that all the points have the same spring attachments (the same type). Remember that these four springs are between mass 1 and mass 0 (ground) and the coordinates are relative to mass 1.

To enter spring attachments between the second mass and the first mass, you need to click "+" to increase the Current Mass No. to 2. Having done that, you need to select the "connected to" combo box. When you click "2", option 1 will appear: 0 and 1, see Figure 3.13. This figure shows how to choose mass number 1 to start creating a new set of spring attachment (between mass 2 and mass 1). As explained previously, mass 2 can connect to mass 1 and mass 0. If we are connecting mass 2 to

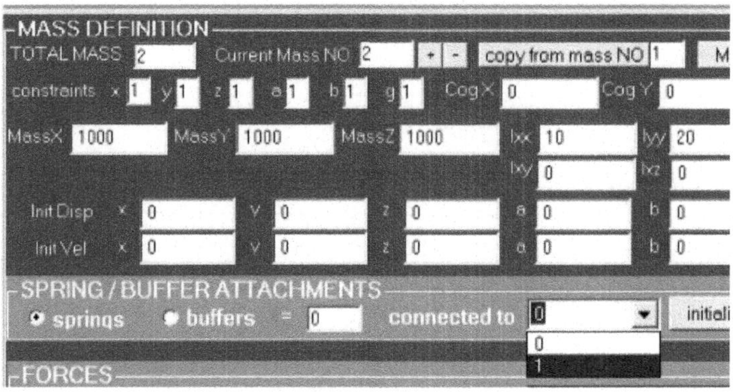

MASS DEFINITION

TOTAL MASS 2   Current Mass NO 2 + - copy from mass NO 1   M

constraints x 1 y 1 z 1 a 1 b 1 g 1   CogX 0   Cog Y 0

MassX 1000   MassY 1000   MassZ 1000   Ixx 10   Iyy 20

Ixy 0   Ixz 0

Init Disp x 0   y 0   z 0   a 0   b 0

Init Vel x 0   y 0   z 0   a 0   b 0

SPRING / BUFFER ATTACHMENTS

⦿ springs   ● buffers = 0   connected to 0 ▼   initiali

0

1

FORCES

FIGURE 3.13   Selecting mass 2 and clicking the "connected to" combo box.

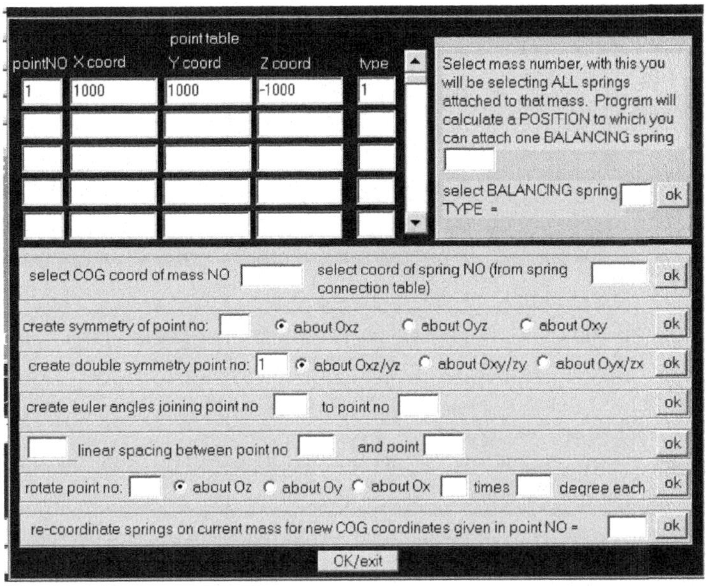

FIGURE 3.14   Use of "spring attachment designer" to create four attachments by double symmetry.

mass 1, then select 1. You may again click the "spring attachment designer" button to create attachment information.

The spring attachment frame will appear with exactly the last set of information (assuming that you have not cleared the text boxes before exiting the frame).

Figure 3.14 shows how to use "spring attachment designer" to create four attachments by double symmetry. Here, click the "ok" button on the "create double symmetry" subframe. Spring attachment information is now generated, between mass 2 and mass 1. Again, coordinates are relative to the current mass 2.

3.6.2.4   Entering Force Data

Make sure that your current mass is 1. When you run VIBRATIO, it assigns a single force of type 1 on mass 1. To see the force data, click "display forces" (on/off). You will see that the "Total Force No." is 1, "current force" is 1, and "type" is also selected to be 1, as shown in Figure 3.15. In this case, "force" is a term that has a wider meaning. Specifically, "flags" on top of "force" magnitudes identify what is meant by "force." Note that "flags" can take values between (and including) 0 and 4. Here, 0 means force does not exist; 1 means ordinary force, measured in N; 2 means prescribed displacement, measured in m; 3 means prescribed velocity, measured in m/sec; and 4 means prescribed acceleration, measured in m/sec^2.

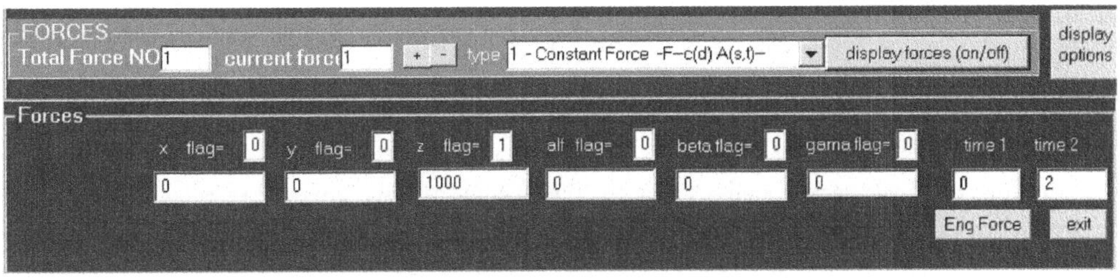

FIGURE 3.15   Default force vector.

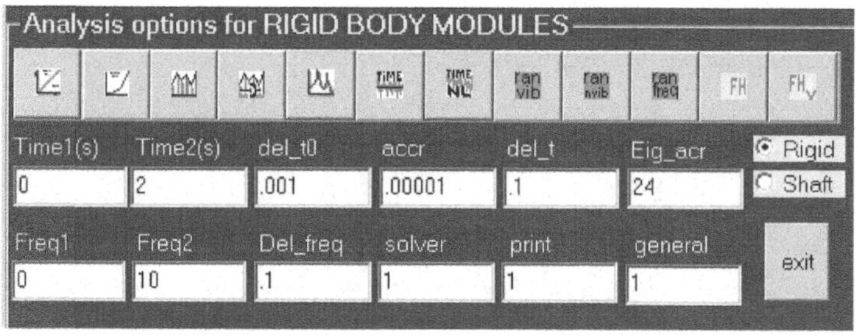

FIGURE 3.16   Analysis options and analysis control parameters.

Note that time 1 and time 2 describe the period in which this force is active. Also, x, y, z, and (α) alpha, (β) beta, (γ) gamma are "force" magnitudes.

Now we have created a two-mass system. Mass 1 is connected to the ground with four springs and mass 2 is connected to mass 1 with four springs. Springs attached to each mass are located at the four corners of the respective masses. There is a constant force acting on mass 1 for 2 sec (although this will be modified according to analysis). Now we are ready to perform analysis.

3.7   Analysis

3.7.1   Analysis Options

To see the analysis frame click the "display analysis options" button.

The analysis option frame as shown in Figure 3.16 will appear. Analysis options offered here are those relevant to the theory presented in this chapter and elsewhere in the book. The theory concerns rigid bodies connected to each other by flexible springs (and dampers). Since flexible shafting analysis is not available in the version provided to the reader, the "rigid" option has to be

FIGURE 3.17   Analysis options available for the provided version of software.

selected. The frame also contains a number of analysis parameters. Only some of the parameters are needed for a given analysis. For example, for deflection, eigenvalue, and eigenvalue (with damping), no parameters are needed. Once you select your analysis, parameters relevant to the analysis will remain visible and anything else will disappear.

Among analysis options, those shown in Figure 3.17 will be functional in your copy of the software. Starting from the left, the first button is for linear deflection, the third button is for eigenvalue analysis (ignores damping) ,the fourth button is for eigenvalue analysis (damping included), the fifth button is for frequency analysis, and the sixth and final button is for time-domain analysis. When you place your cursor on a button, the information box will identify its use.

3.7.2   Eigenvalue Analysis

It is recommended that the first analysis you do is eigenvalue analysis. This will reveal whether you have any inconsistency/error in your data entry and if the created system is physically viable or not.

To perform eigenvalue analysis, click the third button as shown in Figure 3.18. All the analysis parameter data boxes will disappear. The color of the selected button will change to red. This allows the user to enter any appropriate analysis parameter (in this case no parameter is needed).

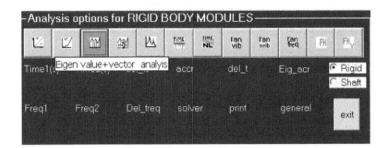

FIGURE 3.18 Eigenvalue analysis button.

FIGURE 3.19 Postprocessing options.

When the button is clicked a second time, the analysis will be performed. To perform eigenvalue analysis for damped system, the procedure is the same but click the fourth button for this analysis.

To see results, you need to click the "display postprocessing options" (on/off) button. The postprocessing options will appear as shown in Figure 3.19. VIBRATIO has a basic text editor that may be used to display textual results. Click the "DISP TEXT" button to view eigenvalue results.

3.7.2.1   Eigenvalue/Vector Analysis Results

The editor window will open, and from the file menu you can then open "eigftext.txt." You should obtain the results given below

X	Y	Z	Alpha	Beta	Gamma
		Frequency in $X = 1.26$ Hz (76 CPM)			
Relative eigenvector values for mass = 1					
1.0000	0.0000	0.0000	0.0000	0.5025	0.0000
Relative eigenvector values for mass = 2					
−0.6180	0.0000	0.0000	0.0000	−0.3106	0.0000
		Frequency in $X = 0.48$ Hz (29 CPM)			
Relative eigenvector values for mass = 1					
0.6180	0.0000	0.0000	0.0000	0.3106	0.0000
Relative eigenvector values for mass = 2					
1.0000	0.0000	0.0000	0.0000	0.5025	0.0000
		Frequency in $Y = 0.48$ Hz (29 CPM)			
Relative eigenvector values for mass = 1					
0.0000	−0.6180	0.0000	0.3098	0.0000	0.0000
Relative eigenvector values for mass = 2					
0.0000	−1.0000	0.0000	0.5012	0.0000	0.0000
		Frequency in $Y = 1.26$ Hz (76 CPM)			
Relative eigenvector values for mass = 1					
0.0000	−1.0000	0.0000	0.5012	0.0000	0.0000
Relative eigenvector values for mass = 2					
0.0000	0.6180	0.0000	−0.3098	0.0000	0.0000

(continued on next page)

(*continued*)

X	Y	Z	Alpha	Beta	Gamma
		Frequency in $Z = 1.78$ Hz (107 CPM)			
Relative eigenvector values for mass = 1					
0.0000	0.0000	1.0000	0.0000	0.0000	0.0000
Relative eigenvector values for mass = 2					
0.0000	0.0000	− 0.6180	0.0000	0.0000	0.0000
		Frequency in $Z = 0.68$ Hz (41 CPM)			
Relative eigenvector values for mass = 1					
0.0000	0.0000	0.6180	0.0000	0.0000	0.0000
Relative eigenvector values for mass = 2					
0.0000	0.0000	1.0000	0.0000	0.0000	0.0000
		Frequency in alpha = 25.26 Hz (1516 CPM)			
Relative eigenvector values for mass = 1					
0.0000	− 0.0050	0.0000	− 1.0000	0.0000	0.0000
Relative eigenvector values for mass = 2					
0.0000	0.0031	0.0000	0.6180	0.0000	0.0000
		Frequency in alpha = 9.65 Hz (579 CPM)			
Relative eigenvector values for mass = 1					
0.0000	− 0.0031	0.0000	− 0.6180	0.0000	0.0000
Relative eigenvector values for mass = 2					
0.0000	− 0.0050	0.0000	− 1.0000	0.0000	0.0000
		Frequency in beta = 17.89 Hz (1073 CPM)			
Relative eigenvector values for mass = 1					
0.0100	0.0000	0.0000	0.0000	− 1.0000	0.0000
Relative eigenvector values for mass = 2					
− 0.0062	0.0000	0.0000	0.0000	0.6180	0.0000
		Frequency in beta = 6.83 Hz (410 CPM)			
Relative eigenvector values for mass = 1					
− 0.0062	0.0000	0.0000	0.0000	0.6180	0.0000
Relative eigenvector values for mass = 2					
− 0.0100	0.0000	0.0000	0.0000	1.0000	0.0000
		Frequency in gamma = 14.57 Hz (874 CPM)			
Relative eigenvector values for mass = 1					
0.0000	0.0000	0.0000	0.0000	0.0000	− 1.0000
Relative eigenvector values for mass = 2					
0.0000	0.0000	0.0000	0.0000	0.0000	0.6180
		Frequency in gamma = 5.56 Hz (334 CPM)			
Relative eigenvector values for mass = 1					
0.0000	0.0000	0.0000	0.0000	0.0000	0.6180
Relative eigenvector values for mass = 2					
0.0000	0.0000	0.0000	0.0000	0.0000	1.0000

This is the result file created as a result of executing eigenvalue analysis (if you had performed eigenvalue analysis with damping you need to open the "eigdtext.txt" file). A successful eigenvalue analysis normally implies physically feasible data (of course, not always).

Now we are ready to perform other analysis options.

3.7.3 Linear Deflection Analysis

The linear deflection equations may be obtained from vibration equations by simply removing the acceleration and velocity terms. If the excitation (force) vector is made of constant values, then the

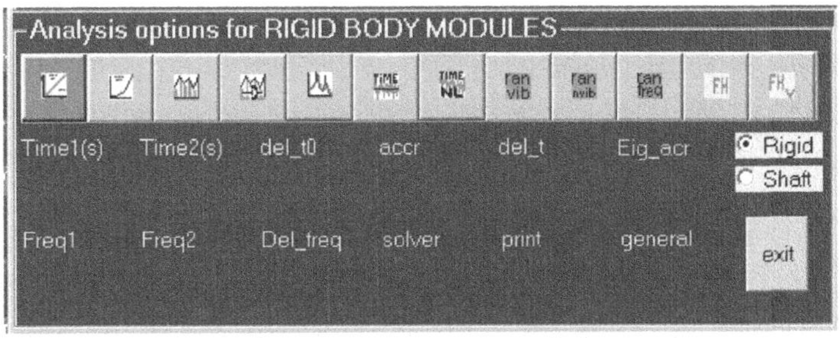

FIGURE 3.20 Linear deflection analysis; no control parameter is needed.

problem is a static deflection problem. By clicking the linear deflection button, you will see that no analysis parameters are needed. Next, click the red button. This will perform a static deflection analysis, as indicated in Figure 3.20.

Numerical results can be displayed using the "DISP TEXT" button from postprocessing options. The deflection results are saved in the "distext.txt" file. This file holds the results given in the next section. Note that, since force is acting on mass 1, mass 2 does not deflect relative to mass 1.

3.7.3.1 Linear Deflection Analysis Results

X	Y	Z	Alpha	Beta	Gamma
		Deflections of mass no. = 1			
0.0000	0.0000	25.0000	0.0000	0.0000	0.0000
		Center of rotation of mass no. = 1			
64,000.0000	64,000.0000	64,000.0000			
		Deflections of mass no. = 2			
0.0000	0.0000	25.0000	0.0000	0.0000	0.0000
		Center of rotation of mass no. = 2			
64,000.0000	64,000.0000	64,000.0000			

Deflection at control points (mm)

Mass no. = 1						
		Position			Deflection	
Point no.	X	Y	Z	x	y	z
Mass no. = 2						
		Position			Deflection	
Point no.	X	Y	Z	x	y	z

Deflections at coupling/mount positions

Mount no.	Position			Deflection		
	X	Y	Z	x	y	z
Mass no. = 1						
1	1000.000	1000.000	− 1000.000	0.000	0.000	25.000
2	− 1000.000	1000.000	− 1000.000	0.000	0.000	25.000
3	− 1000.000	− 1000.000	− 1000.000	0.000	0.000	25.000
4	1000.000	− 1000.000	− 1000.000	0.000	0.000	25.000

(*continued on next page*)

(*continued*)

Mount no.	Position			Deflection		
	X	Y	Z	x	y	z
Mass no. = 2						
5	1000.000	1000.000	− 1000.000	0.000	0.000	25.000
6	− 1000.000	1000.000	− 1000.000	0.000	0.000	25.000
7	− 1000.000	− 1000.000	− 1000.000	0.000	0.000	25.000
8	1000.000	− 1000.000	− 1000.000	0.000	0.000	25.000

Global and Local Deflections of Couplings/Mounts

Coordinates mount no.	Global Deflection			Local Deflection		
	xp	yp	zp	X	Y	Z
Mass no. = 1						
1	0.00	0.00	25.00	0.00	0.00	25.00
2	0.00	0.00	25.00	0.00	0.00	25.00
3	0.00	0.00	25.00	0.00	0.00	25.00
4	0.00	0.00	25.00	0.00	0.00	25.00
Mass no = 2						
5	0.00	0.00	0.00	0.00	0.00	0.00
6	0.00	0.00	0.00	0.00	0.00	0.00
7	0.00	0.00	0.00	0.00	0.00	0.00
8	0.00	0.00	0.00	0.00	0.00	0.00

3.7.4 Frequency Analysis

In order to perform a frequency analysis, the force selection must be relevant to the frequency analysis. Force options between 16 and 21 are for frequency analysis. Option 16 ($A \sin Wt$) is chosen to demonstrate the frequency analysis, as shown in Figure 3.21.

Here, default excitation is in the z direction. Flag = 1 means this is an ordinary sinusoidal force with 1000 N excitation amplitude. It exists in the frequency range from 0 to 2 Hz (freq 1 and freq 2) acting in

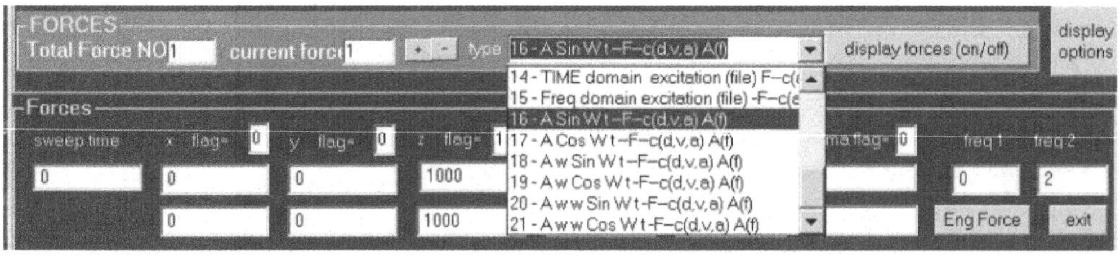

FIGURE 3.21 Frequency analysis options 16 to 21.

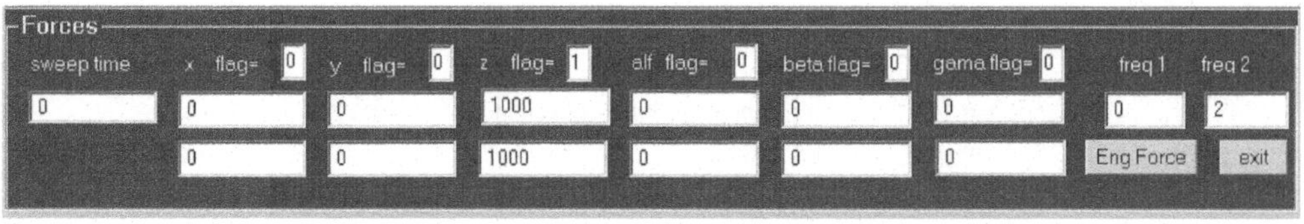

FIGURE 3.22 Constant force amplitude between 0 and 2 Hz.

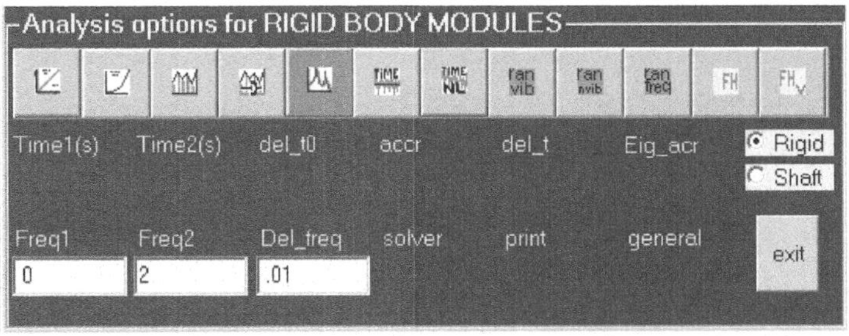

FIGURE 3.23 Frequency analysis parameters.

the z direction, as shown in Figure 3.22. You may have more than one force, with different amplitudes and different frequency ranges. Sweep time, in seconds (any number other than zero), refers to the time taken to sweep from freq 1 to freq 2. If the sweep time is infinitely slow, then a zero entry signifies this (exactly the opposite meaning!). There are two amplitudes for each direction, the top entry refers to the amplitude at freq 1 and the bottom one refers to the amplitude at freq 2. Amplitudes between these frequencies vary linearly.

To execute a frequency analysis, select analysis options and click the button shown in Figure 3.23. Frequency analysis parameters: start frequency, end frequency, and frequency steps are shown in this figure.

Specifically, the options Freq 1 and Freq 2 will appear here. These relate to the frequency range to be analyzed. Del_freq is the frequency step. Your choice of frequency range should not be arbitrary. Either the problem dictates the range or you may be interested in amplitudes of the system at resonant frequencies. In the latter case, you may look at the eigenvalue results to identify which resonant frequencies will be exhibited in the selected range. For the example considered, the excitation force is in the z direction. Therefore, one needs to find out the resonant frequencies in the z direction (or coupled modes which include the z direction). If the eigenvalue results are studied, one can see that the resonant frequencies in the z direction are given as: 0.68 and 1.78 Hz. Therefore, analysis beyond 2 Hz in the z direction will not reveal any other resonances. Clicking the red button a second time will execute the analysis. To see the results you need to click the "display postprocessing" button, and from the postprocessing frame click the "plot freq" button. The frequency-plotting window will appear. The frequency motion in six directions will be plotted for mass 1. "+" can be used to move to the next mass. You may also display motion curves of all bodies together by choosing the "plot all" option. You will end up with the results shown in Figure 3.24.

3.7.5 Time-Domain Analysis

Forces 1 to 14 are for the time domain, although 13 and 14 are not available in this version of VIBRATIO. When option 1 is selected in the force "type" option, the default force would be 1000 N in the z direction, remaining active between 0 and 2 sec. Modify Time 2 to 0.01 sec (10 msec). You can use option 1 for a rectangular shock but there are also other shock options. Figure 3.25 shows a constant force, magnitude 10,000 N acting in the z direction between 0 and 0.1 sec.

To perform a time domain analysis, the button shown in Figure 3.26 needs to be clicked. Relevant analysis parameters will appear as shown in this figure. Time 1 and Time 2 give the range of the time domain analysis. If Time 1 is chosen not to start from 0, even though the resulting data will not be collected until Time 1 reaches, still the solution will be executed starting from start time $= 0$. In other words, if Time 1 is not zero then this will be the time when data recording starts. Analysis itself, irrespective of Time 1, always starts from 0 sec. del_t0 is the initial step length for integration

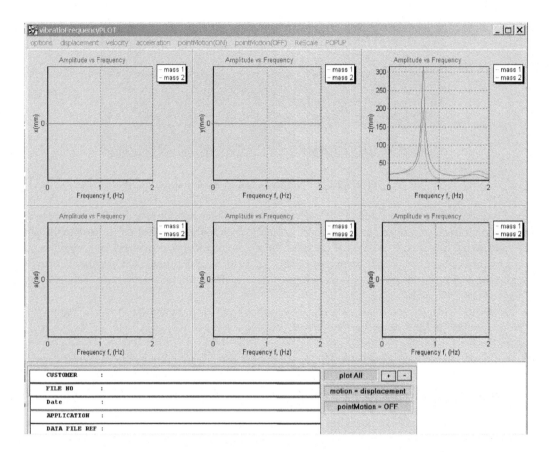

FIGURE 3.24 Frequency analysis results.

(The time-domain analysis uses variable step length Runge–Kutta and the time integration step where an algorithm automatically modifies the integration step to achieve the required accuracy). Accr is the accuracy of integration and del_t is the time step at which data is sampled. To execute, click the button again.

3.7.5.1   Time-Domain Results

Time domain results will be displayed by clicking the plot-time button on the postprocessing frame. The results shown in Figure 3.27 will be obtained, again from the "options" menu, when "plot all" is selected.

FIGURE 3.25 Constant force representation.

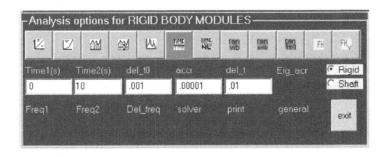

FIGURE 3.26 Time-domain analysis button and analysis control parameters.

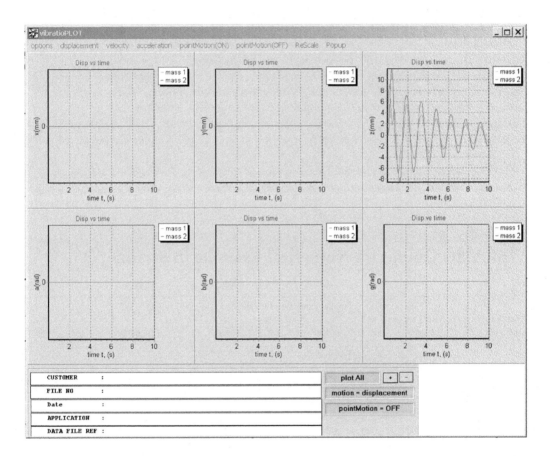

FIGURE 3.27 Time-domain results window.

3.8 Comments

This chapter presented a method of analyzing general multiple rigid-body systems interconnected by linear springs and linear dampers. The mathematical modeling was presented for small vibrations where nonlinear geometry effects and gyroscopic couplings were negligible and the deflection characteristics of the mountings were linear. There is a shortage of published material on general mathematical modeling of flexibly supported rigid multibody systems for vibration analysis. Detailed mathematics of time domain, frequency domain and eigenvalue/eigenvector analyses can be found in standard vibration textbooks. Some are listed in the references section, but this list is not exhaustive. This chapter also gives a precise and clear formulation suitable for computational implementation. The formulation presented in this chapter forms the core of the vibration analysis suite "VIBRATIO", a version of which is made available to the users of this handbook, at www.signal-research.com, as indicated in the text.

Bibliography

Bishop, R.F.D. and Johnson, D.C. 1960. *The Mechanics of Vibration*, Cambridge University Press, New York.

Caughey, T.K., Classical normal modes in damped linear dynamic systems, *J. Appl. Mech.*, 27, 269–271, 1960.

den Hartog, J.P. 1956. *Mechanical Vibration*, 4th ed., McGraw-Hill, New York.

de Silva, C.W. 2006. *VIBRATION: Fundamentals and Practice*, 2nd Edition, Taylor & Francis, CRC Press, Boca Raton, FL.

Jacobsen, L.S. and Ayre, R.S. 1958. *Engineering Vibrations*, McGraw-Hill, New York.

Meirovitch, L. 1986. *Elements of Vibration Analysis*, 2nd ed., McGraw-Hill, New York.

Meirovitch, L. 1967. *Analytical Methods in Vibrations*, Macmillan, New York.

Ralston, A. 1965. *A First Course in Numerical Analysis*, McGraw-Hill, New York.

Thomson, W.T. 1988. *Theory of Vibrations with Applications*, 3rd ed., Prentice Hall, Englewood Cliffs, NJ.

Timoshenko, S., Young, D.H., and Weaver, W. Jr. 1974. *Vibration Problems in Engineering*, 4th ed., Wiley, New York.

VIBRATIO, information and download for the readers of this chapter, www.signal-research.com.

Wilkinson, J.H. 1965. *The Algebraic Eigenvalue Problem*, Clarendon Press, Oxford.

Appendix 3A
VIBRATIO Output for Numerical Example in Section 3.3

X	Y	Z	Alpha	Beta	Gamma
		Frequency in $X = 1.01$ Hz (60 CPM)			
Relative eigenvector values for mass = 1					
1.0000	0.0000	0.0000	0.0000	0.0000	0.0000
		Frequency in $Y = 1.42$ Hz (85 CPM)			
Relative eigenvector values for mass = 1					
0.0000	1.0000	0.0000	0.0000	0.0000	0.0000
		Frequency in $Z = 1.74$ Hz (105 CPM)			
Relative eigenvector values for mass = 1					
0.0000	0.0000	1.0000	0.0000	0.0000	0.0000
		Frequency in alpha = 34.87 Hz (2092 CPM)			
Relative eigenvector values for mass = 1					
0.0000	0.0000	0.0000	1.0000	0.0000	0.0000
		Frequency in beta = 12.33 Hz (740 CPM)			
Relative eigenvector values for mass = 1					
0.0000	0.0000	0.0000	0.0000	1.0000	0.0000
		Frequency in gamma = 14.24 Hz (854 CPM)			
Relative eigenvector values for mass = 1					
0.0000	0.0000	0.0000	0.0000	0.0000	1.0000

4

Finite Element Applications in Dynamics

Mohamed S. Gadala
The University of British Columbia

Summary

This chapter discusses the use of the finite element (FE) method in problems of vibrations and structural dynamics. The first three sections outline the main steps in modeling a physical problem for a specific dynamic analysis. Section 4.1 concentrates on the finite element aspect of the modeling process. The basis for geometric modeling is outlined and an overview of commonly used types of elements in typical commercial programs is presented. A summary of element capabilities is given at the end of the section (Table 4.1). Section 4.2 discusses the basis for classifying and choosing a particular type of dynamic analysis, including modal, harmonic or frequency response, transient and shock, and random analysis. Section 4.3 discusses some special aspects in modeling that are pertinent to dynamic system analysis; namely, choice of master and slave degrees of freedom (DoF), lumped vs. consistent mass modeling, and use of symmetry in dynamic analysis.

The second part of the chapter discusses solution methods and damping considerations, and outlines basic steps for performing various dynamic analyses. Section 4.4 briefly presents the theory and equations for various dynamic analyses. The analysis types included in the discussion are direct integration and modal superposition. In the direct integration analysis, both implicit and explicit schemes are discussed and an example of each is presented. Most of the theory and equations are presented in a summarized form without rigorous mathematical proofs. Section 4.5 describes details of various dynamic analyses and provides a brief discussion of the choice of the solution method for each analysis. Emphasis is placed on the basic steps required to perform a particular analysis type. Modal analysis, transient analysis (direct integration and mode superposition approaches), frequency response harmonic analysis, and random response analysis are presented. The various methods of combination of modal responses are also discussed. Finally, Section 4.6 provides some general guidelines for a typical dynamic analysis using the FE method.

4.1   Problem and Element Classification

The first step in any finite element (FE) analysis is to build a model of the physical structure to be analyzed. This is an important step and normally requires extensive time and interaction with the analyst. The adequacy of the model, assumptions involved, and types of elements used for a specific structure and analysis type establish the accuracy level of the FE analysis. Much research effort is being devoted to automating this process by providing options for automatic mesh generation, automatic mesh refinement, error estimation, and error bounds. Such research has resulted in a significant reduction in the time needed for this step but the role of the analyst is still dominant and vital in obtaining a good mesh and model for the problem.

The modeling process can be divided into steps as follows:

- Build a geometric database for the structure. This includes description of the characteristic geometric features of the structure, such as boundaries, holes, intersections, curvatures, etc., in the finite element program database. The level of geometric detail has an important effect on the accuracy and the size of the model.
- Build a FE model for the geometric model. This may include important aspects such as:
  ○ Establishing the type of analysis to be performed,
  ○ Choosing the appropriate element or elements for building the model,
  ○ Considering aspects of symmetry in the structure, and
  ○ Establishing critical areas for increasing mesh density.
- Apply constraints and loading boundary conditions.
- Establish the material model(s) to be used in the analysis.
- Perform various options of model checks.
- Solve sample load cases and compare the results with hand calculations or experimental results in order to check the behavior and response of the model.
- Fine tune the model based on the results obtained from sample load cases.

Most of the time, the above steps are rather linked together, and an overall knowledge of the problem is required to perform a specific step. An important decision that should be made at the beginning of the analysis is to identify the category of the problem. This will have an impact on the first three steps mentioned above. In the first part of the present section, we provide a brief discussion on the geometric modeling aspects of the problem. The rest of the section then provides general guidelines on classifying the problem into one of the main categories available in typical commercial finite element programs; namely, truss, beam, two-dimensional, shell, or three-dimensional problems. We also provide simple examples for each type of problem. The concepts of element choice and problem classification are summarized in a table at the end of the section.

4.1.1   Geometric Modeling

Geometric modeling simply means transforming a physical problem into a geometric database in a FE program. The process is very similar to creating an engineering drawing or a model in a CAD program. Most FE programs have built-in preprocessors that are dedicated to generating the geometry database. Generally, FE preprocessors have similar capabilities to those available in CAD programs. In many cases an existing geometric database may be available for the structure in a CAD program and may be imported into the FE program database. However, there can be translation problems between the two databases, especially in three-dimensional and shell structures. In most cases, it is faster to regenerate the geometric database for the problem directly using the FE program.

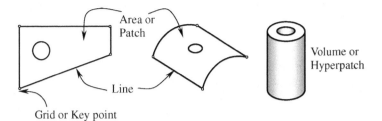

FIGURE 4.1 Geometric modeling entities.

Finite element programs use specific building blocks or entities to build the geometry of a structure (Kamel, 1991; *NISA User's Manual*, 1992; *ANSYS User's Manual*, 2003). The theory on which geometry generation is based is quite simple and basically relies on parametric cubic modeling of curves and surfaces in space. Figure 4.1 shows the basic entities used by most programs for building a geometric model. We define these basic geometric entities and briefly discuss their use in practical modeling of structures.

4.1.1.1 Key Point

A key point is a coordinate location in space. In two-dimensional space, a key point is uniquely defined by two coordinates (e.g., x, y; r, θ). In three-dimensional space, a key point is uniquely defined by three coordinates (e.g., x, y, z; r, θ, z). Key points are normally the starting building blocks in the geometry. Connection between key points will generate lines, surfaces, or volumes. On the other hand, most programs will be capable of extracting key points from the end points or corners of lines, surfaces, and volumes. Key points should not be confused with nodes, and should be considered only as spatial locations in space. One may place a node at the same location as a key point or leave the location without node generation.

4.1.1.2 Line or Line Segment

A line or line segment is a portion of a cubic spline curve bounded on both ends by a key point. A line may be straight or curved. The curvature of a line is limited only by its parametric cubic equation in the program. If a physical curve in the structure is presented by up to and including a third-order parametric cubic equation, it may be modeled exactly by a single line. In many practical situations, however, this is not the case. As an example, the parametric equation of a circle is not cubic and the recommended way to model a circle accurately is to break it into at least four lines. Breaking a space curve into many lines will always increase the accuracy of the geometric modeling but entails increasing the complexity of the model. The analyst should be careful in situations where the order of the line to be modeled is not known. A common case is the generation of the line of intersection between two surfaces. In such situations, it is important to break the intersection line into a few separate lines.

Most programs provide extensive methods for line generation. These may include generation by joining two grid points, cubic spline fitting of four grid points, best fitting a curve between several grid points, extracting the edges of a surface or a volume, intersection of surfaces, and mirroring and copying other lines.

4.1.1.3 Area or Patch

An area or a patch is a portion of a bi-cubic surface completely bounded by three line segments (for triangular areas) or four line segments (for quadrilateral areas). The same limitations discussed above for using line segments may be extended here by realizing that the area is modeled by parametric cubic equations in two directions representing the two edges of the area.

Most programs provide extensive methods for generating areas or patches. These include generation by sweeping the space between two lines, filling in the area between four edge lines, rotating a line about an axis, extracting the boundary surfaces of a volume, and intersection between volumes.

4.1.1.4 Volume or Hyperpatch

A volume or a hyperpatch is a portion of tri-cubic solid completely bounded by four areas (for tetrahedron volumes) or six areas (for brick volumes). The equations used to model the edge lines of a volume are still parametric cubic equations and the same restrictions discussed for lines can be extended to a volume in the three directions of the volume.

As for line and area generation, most programs provide a wide variety of methods for generating volumes. These may include generation by sweeping the volume between two surfaces, filling the volume between bounding surfaces, rotating an area about an axis, and copying and mirror imaging the existing volume.

An important addition in the generation of volumes is the ability of many programs to use solid primitives as building blocks (Kamel, 1991; *ANSYS User's Manual*, 2003). These solid primitives may include tetrahedrons, cubes, cylinders, conical volumes, spheres, torus elements, and other standard volumes. These may be used as building blocks that may be combined, subtracted, or intersected with each other. Many programs provide simple Boolean operations to use for such processes.

REMARKS

- In most modeling cases, it is faster to regenerate the geometric database of the problem directly using the FE program, especially in complicated and three-dimensional models.
- The basic building blocks in geometric modeling are key points, lines, areas, and volumes.
- Many programs provide the ability to create solid primitives as building blocks. These solid primitives may include tetrahedrons, cubes, cylinders, conical volumes, spheres, toruses, and other standard volumes.
- Boolean operations are normally used to combine, subtract, or intersect various geometric entities.

4.1.2 Discrete Element Types in FE Programs

Most commercial FE programs have extensive element libraries that may be used in static and dynamic analyses. For dynamic analysis, it may be convenient to classify elements into discrete and continuum types. Discrete types include concentrated (lumped) mass and inertia, spring, and damper elements, whereas continuum (distributed) types include all other one-dimensional (1-D), two-dimensional (2-D), and three-dimensional (3-D) deformable elements. In this section, we briefly discuss the discrete type of elements whereas the continuum type will be discussed in detail in subsequent sections.

4.1.2.1 Concentrated Mass/Inertia Element

A concentrated mass/inertia element represents a structural mass and moment of inertia concentrated at one point and has six DoF: three translational and three rotational. The mass and rotary (moment of) inertia may be assigned different values in the three coordinate directions (see Figure 4.2), even though, typically, the mass is the same in all three directions (see Chapter 3). The element is rigid with no geometrical properties and it only contributes to the global mass matrix of the structure. In building up a model, the element may be attached to a structural node of other deformable or elastic elements or be positioned in space and attached to structure nodes through rigid elements or elastic spring and/or damper elements. Most FE programs provide rotary inertia quantities for various components of the geometric model as part of the standard preprocessing data. These can be used to model parts of the structure as lumped mass and inertia that may be connected to the structure through elastic elements.

An example of the use of a concentrated mass and inertia element is the modeling of a heavy, rigid machine mounted on an elastic support. The mass and inertia effects of the machine may be represented by a concentrated or lumped mass/inertia element at the center of the machine. Care must be taken in connecting this element to the deformable elements of the structure or the support. In general, a rigid link may be used to connect the mass/inertia element to the nearest node of a deformable element rather than placing the mass/inertia element directly on that node (Cook et al., 1989). This will generally account for proper interaction between translational and rotational DoF of the mass and the structure.

It should be noted that some computational algorithms might have problems with zero diagonal values of the mass matrix. This happens if the inertia terms of the mass/inertia element are assigned as zero. This can be avoided by always assigning an arbitrary and small value to the rotary inertia terms.

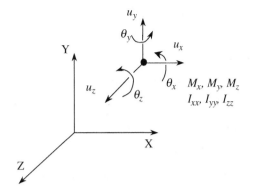

FIGURE 4.2 Three-dimensional mass/inertia element.

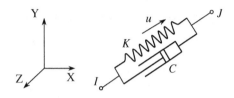

FIGURE 4.3 One-dimensional spring/damper element.

4.1.2.2 Spring and Damper Elements

Most programs provide 1-D spring and damper elements that can be used to model hydraulic cylinders, discrete dampers, and shock absorbers. The element normally has one spring constant and one damping coefficient (along the element axis defined by the nodes *I* and *J* as shown in Figure 4.3). It is easy to model 3-D stiffness and damping characteristics by replicating the element in the required directions. A mass may be attached to one or two nodes of the element. The force transmitted through the element nodes is the sum of the spring and the damping forces along the element axis.

REMARKS

- A concentrated mass/inertia element may be used to model lumped mass/inertia at specific points in the structure. Likewise, spring or damper elements may be used to model stiffness and damping characteristics between two points in the structure.
- A concentrated mass/inertia element represents a structural mass and inertia at a point and has six DoF: three translational and three rotational. Different mass and inertia values may be assigned in different directions.
- Three-dimensional stiffness and damping characteristics between two nodes may be modeled by replicating 1-D spring or damper elements in the required directions.

4.1.3 Truss and Beam Problems

Three-dimensional truss and beam elements are shown in Figure 4.4. The main difference between the two elements is the fact that beams may support bending moments and have rotations as extra DoF at

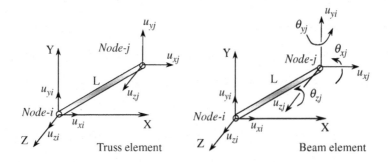

FIGURE 4.4 Three-dimensional truss and beam elements. (Nodal rotations for the beam element are only shown for node-*j*.)

each node. This means that a beam element may have loads along the beam axis and not necessary only at the end points.

The following conditions should be satisfied for a structure to be classified as a truss or a beam:

- *Geometry*:
  - Thin, slender, straight bars or rods with pin joints at both ends. Depending on the application, the element may be considered as a truss, a link, a cable, a spring, etc.

    If the joints are fixed or the bar is curved, the problem will be classified as a beam type. Fixed joints may be created by completely fixing the beam ends to a wall or support, or to another member. This is normally realized if the joint is built-in, welded, or fully bolted to the support or to other members.
  - Geometry may be 1-D, 2-D, or 3-D.
- *Loading*:
  - For truss problems, loading may only be by tension or compression of the members. This may be achieved by having only concentrated loading at the joints (no bending moment) and no loadings along the member.

    For beam problems, loading may be in any direction and may be applied along the member axis. This will generally create a bending moment and the member should have rotations as DoF.
  - Thermal loading may also be applied to the member.
  - Body or inertia forces will generally create a bending moment and violate the truss condition and so may be applied only to beam problems. Such loads may be applied, however, to truss elements under the simplification of assuming the inertia effects to be applied only at the nodes or the joints and ignoring the bending moment that will be created on the member.

Degrees of freedom: For 3-D truss elements, the DoF per node are the displacement in the three coordinate directions: u_x, u_y, and u_z. Two-dimensional truss members only have u_x and u_y as DoF. Three-dimensional beam elements have all six DoF: three translational, u_x, u_y, and u_z, as well as three rotational, θ_x, θ_y, and θ_z. Two-dimensional beam elements have u_x, u_y, and θ_z as DoF.

Element shapes: Various element shapes commonly available in commercial finite element programs are shown in Figure 4.5. The cross section of the element may be a solid or hollow prismatic section, e.g., a rectangular, circular, trapezoidal, or thin-walled section or a channel, thin-walled tubular, I-section, etc.

2-node (linear)	3-node (quadratic)	4-node (cubic)
Truss and beam	Beam only	Beam only

FIGURE 4.5 Various element shapes for truss and beam elements.

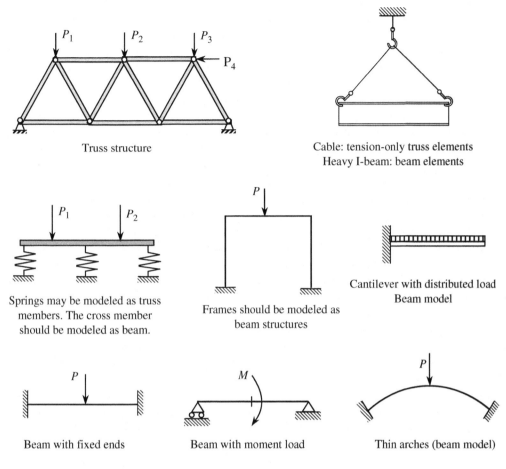

FIGURE 4.6 Typical applications for truss and beam elements.

The analyst will generally be required to identify the orientation of the beam cross section by specifying local axes or identifying key points for the program to locate the local axes. Most programs are capable of modeling variable cross section beam elements to avoid excessive subdivision of the beam into smaller elements.

Example applications: Figure 4.6 shows sample applications for the use of truss and beam elements.

REMARKS

- Truss elements have translational DoF whereas beam elements have both translational and rotational DoF at the nodes. Three-dimensional truss elements have 3 DoF/node (u_x, u_y, and u_z), whereas 3-D beam elements have 6 DoF/node (u_x, u_y, u_z, θ_x, θ_y, and θ_z).
- Truss structures may only carry loads at the end of each truss whereas beam structures may also carry loads along the beam length.
- Curved members should be modeled with beam elements.
- Problems with body and inertia forces should generally be modeled as beam elements. If lumping of the body forces is assumed and the effect of the bending moment created by body forces is neglected, the effect may be modeled with truss elements.

4.1.4   Two-Dimensional Problems

4.1.4.1   Two-Dimensional Plane Stress Problems

Figure 4.7 shows a typical structure, which may be considered as a 2-D plane stress problem.

The following geometry and load conditions should be satisfied for a structure to be categorized as a 2-D plane stress structure (Boresi and Chong, 2000).

Conditions:
- *Geometry*:
  - A flat, thin surface in one plane (e.g., *xy*-plane), simply connected or multiply connected with a small thickness in the third direction (e.g., *z*-direction).
  - The boundary of the structure in the *xy*-plane may be straight or curved.
- *Loading*:
  - Loading is restricted to the plane of the structure (e.g., *xy*-plane).
  - Thermal loading may also be applied to the plane (i.e., $T = T(x, y)$).
  - Body or inertia forces may be due to linear or angular acceleration in the plane of the structure.

Degrees of freedom: For 2-D elements, the DoF per node are the displacements in the two coordinate directions, u_x and u_y.

Typical element shapes (see Figure 4.8): Elements may be triangular or quadrilateral. Most programs provide the option of having 3, 6, or 9 nodes for triangular elements and 4, 8, or 12 nodes for quadrilateral elements. Increasing the number of nodes will increase the element accuracy at the expense of having more DoF and requiring more CPU time to solve the problem. This is normally called a "p-conversion" approach in finite elements. The same goal may be achieved by increasing the number of elements while fixing the number of nodes per element, which is usually called "h-conversion" (Zeng et al., 1992; Babuska and Guo, 1996). The choice of method to obtain greater accuracy is rather arbitrary and mostly depends on the availability of the option in the program. Only a limited number of programs provide extensive "p-conversion" options.

Example applications: Figure 4.9 shows some typical examples that may be modeled using a plane stress assumption. It should be noted such models cannot capture any out-of-plane modes of vibration. If such vibration modes are of concern, the model should be capable of capturing lateral displacement DoF and out-of-plane rotations. This may be realized by using shell or 3-D solid elements as will be discussed in the following sections.

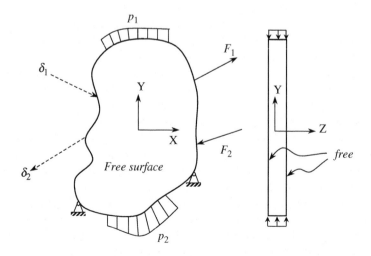

FIGURE 4.7   A general 2-D plane stress structure.

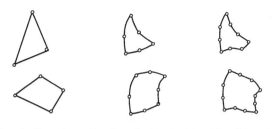

3 or 4-node (linear)   6 or 8-node (quadratic)   9 or 12-node (cubic)

FIGURE 4.8 Typical plane stress elements.

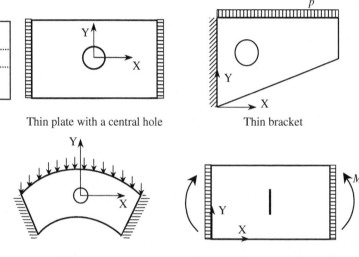

Thin plate with a central hole   Thin bracket

Thin arch   Through thickness crack in a plate

FIGURE 4.9 Typical plane stress examples.

4.1.4.2 Two-Dimensional Plane Strain Problems

Figure 4.10 shows a typical structure that may be considered as a 2-D plane strain.

The following geometry and load conditions should be satisfied for a structure to be categorized as a plane strain structure (Boresi and Chong, 2000).

Conditions:
- *Geometry*:
  - A flat surface in a plane (e.g., *xy*-plane), simply or multiply connected with large thickness in the third or *z*-direction (with much larger dimensions than in the *xy*-plane).
  - The boundary of the structure in the *xy*-plane may be straight or curved.
- *Loading*:
  - Loading is restricted to the plane of the structure (e.g., *xy*-plane) with possible uniform loading or constraint in the *z*-direction. The loading in the *z*-direction will remain only as a function of the *x* and *y* coordinates; i.e., $P_z = P_z(x, y)$.
  - Thermal loading may also be applied to the plane (i.e., $T = T(x, y)$).
  - Body or inertia forces may be applied to the plane.

Degrees of freedom: For 2-D plane strain elements, the DoF per node are the displacement in the two coordinate directions u_x and u_y. The plane strain assumption means that the strain in *z*-direction will be zero and that the stress will be nonzero, $\varepsilon_z = 0$ and $\sigma_z \neq 0$.

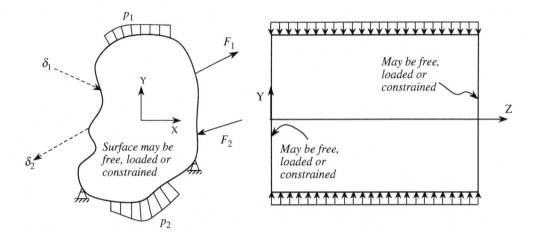

FIGURE 4.10   A general 2-D plane stress structure.

Typical element shapes: The same element shapes as in the case of plane stress are available in most commercial FE programs (see Figure 4.8). The comment above on p-conversion and h-conversion also applies here.

Example applications: Figure 4.11 shows some typical examples that may be modeled using a plane strain assumption. Once again, it should be noted that such a model cannot caputre any out-of-plane modes of vibration. If such vibration modes are of concern, the model should be capable of capturing lateral displacement DoF and out-of-plane rotations. This may be realized by using shell or 3-D solid elements as will be discussed in the following sections.

4.1.4.3   Two-Dimensional Axisymmetric Problems

Axisymmetric problems are characterized by having an axis of symmetry or axis of revolution for geometry and loading. Referring to Figure 4.12, any arbitrary plane that passes by the axis of symmetry will be a plane of symmetry. Symmetry means that the two halves of the structure on each side of the plane of symmetry are mirror images of each other. Symmetry must be satisfied for all aspects affecting the response of the structure including geometry, load, constraint, and material properties.

The following conditions summarize the requirements for categorizing a problem as axisymmetric (Boresi and Chong, 2000).

Conditions:
- *Geometry*:
  - A solid of revolution formed by rotating a flat area (e.g., in the rz-plane) around an axis of symmetry (e.g., the z-axis).
  - The boundary of the flat area in the rz-plane may be straight or curved and the area may be simply or multiply connected.
- *Loading*:
  - Loading is restricted to the rz-plane with no variation of loading in the θ-direction. No loading in the θ-direction.
  - Thermal loading may be applied in the rz-plane (i.e., $T = T(r, z)$).
  - Body or inertia forces may be applied in the rz-plane.

Degrees of freedom: For 2-D-axisymmetric elements, the DoF per node are the displacements in the two coordinate directions u_r and u_z. Some programs use x and y coordinates to replace the r and z axes, respectively.

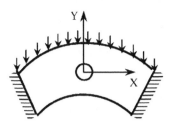
Very thick arch

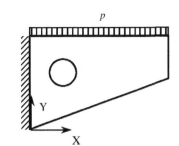
Thick cantilever or bracket support

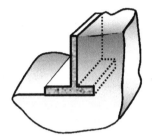
Retaining wall

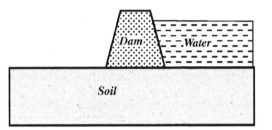
Impermeable earth dam and its base

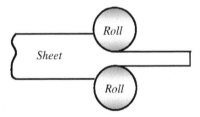
Sheet metal rolling

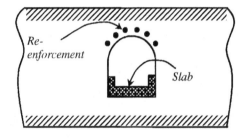
Subway tunnel

FIGURE 4.11   Typical plane strain examples.

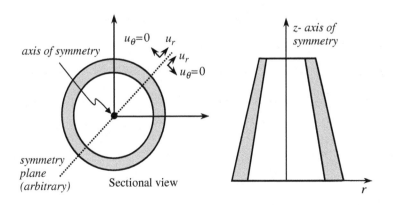

FIGURE 4.12   Axial symmetry conditions.

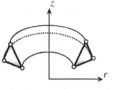

| Section in an axisymmetric | 3 or 4-node (linear) | 6 or 8-node (quadratic) | 9 or 12-node (cubic) |

FIGURE 4.13 Typical axisymmetric elements.

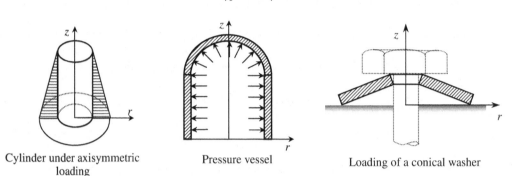

Cylinder under axisymmetric loading   Pressure vessel   Loading of a conical washer

FIGURE 4.14 Typical axisymmetric examples.

Typical element shapes: Figure 4.13 shows typical axisymmetric element shapes that are commonly available in commercial FE programs. The elements are shaped as a torus with various cross sections as shown in the figure.

Application examples: Figure 4.14 shows typical axisymmetric examples.

4.1.5 Shell and Plate Problems

Shell and plate problems are quite similar to plane stress problems (see Figure 4.15). We first recall the conditions for a plane stress problem: (1) that the geometry has to be flat in one plane with a small thickness, and (2) that the load has to be in the same plane. If either of these two conditions is violated, the problem becomes a shell problem. For example, if the load is a moment about any direction other than the z-direction then it has an out-of-plane component, or if the geometry is not flat then the problem ceases to be a plane stress problem and should be modeled as a shell problem. Normally, shell structures would still maintain a small thickness in the direction normal to the surface of the shell. This maintains the condition that the stress normal to the shell surface will be zero, although there may still be a pressure applied to the surface of the shell.

The following summarizes the conditions for a shell structure.

- *Geometry*:
  - Thin surfaces or plates that have a thickness in the direction normal to the surface. The midsurface or midplane of the structure may be flat or curved.
  - The boundary or edges of the structure may be straight or curved.
- *Loading*:
  - Both in-plane and out-of-plane loadings are permitted.
  - Thermal loading may also be applied in all x-, y-, and z-directions.
  - Body or inertia forces may be due to linear or angular acceleration in all three directions.

Degrees of freedom: Three-dimensional shell elements have six DoF: three translational, u_x, u_y, and u_z, and three global rotational, θ_x, θ_y, and θ_z.

REMARKS

- Structures with a flat planar surface of a small thickness and with loading only in the plane (e.g., *xy*) of the structure (no out-of-plane loading or constraints) are categorized as plane stress structures. Such structures will have three nonzero stress components: σ_{xx}, σ_{yy}, and t_{xy}.
- Structures with a flat planar surface (e.g., *xy*) but a very large dimension in the third direction and with loading only in the plane of the structure (with possible uniform loading or constraints in the third direction) are categorized as plane strain structures. Such structures will have four nonzero stress components: σ_{xx}, σ_{yy}, σ_{zz}, and t_{xy}.
- Axisymmetric problems are characterized by having an axis of symmetry or axis of revolution (e.g., *z*-axis) for geometry and loading. Such structures will have four nonzero stress components: σ_{rr}, $\sigma_{\theta\theta}$, σ_{zz}, and t_{rz}.
- Two-dimensional problems may be modeled by elements having two translational DoF per node. The elements may be triangular with 3, 6, or 9 nodes or quadrilateral with 4, 8, or 12 nodes. Axisymmetric elements are shaped as a torus with triangular or quadrilateral cross-sections.
- Increasing the number of elements while fixing the nodes per element is called "h-conversion," whereas increasing the number of nodes per element while fixing the number of elements is called "p-conversion."
- Two-dimensional models cannot capture out-of-plane vibration modes. If such vibration modes are of concern, the model should be capable of capturing lateral displacement DoF and out-of-plane rotations.

Element shapes: Figure 4.16 shows some typical shell elements available in most commercial FE programs. To properly model a curved shell with linear elements that are flat requires using a large number of elements to capture the curvature. Quadratic and cubic elements, on the other hand, may have curvature in two directions and a much reduced number of elements will be needed to model a curved shell. Most programs are capable of modeling variable thickness shell and plate elements.

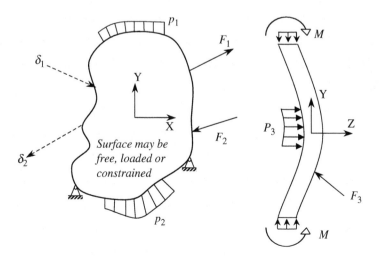

FIGURE 4.15 Shell and plate structures.

Some programs offer shell elements that only have membrane capabilities and others that have both membrane and bending capabilities.

In shell and plate analyses, it is important to note that linear elements, i.e., three-node triangles and four-node quadrilaterals, may behave in an unrealistically stiff manner in shear deformation when the element thickness to size ratio is very small. This phenomenon is normally called "shear locking" (Bathe and Dvorkin, 1985; Bathe, 1996; Luo, 1998; Cesar de Sa et al., 2002). The problem occurs when in-plane displacements are coupled with section rotations in the governing equation of the element and when low-order interpolations (linear elements) are adopted. The same phenomenon is also evident in beam elements. Similarly, when in-plane displacements are coupled with section rotations in the governing equations and low-order interpolations are used, "membrane locking" will be evident.

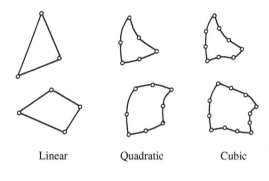

Linear   Quadratic   Cubic

FIGURE 4.16   Typical shell elements.

Most programs provide various remedies for shear- and membrane-locking problems. These include selective/reduced integration, enhanced assumed strain method, mixed field method, etc. Depending on the availability of a particular method in the program, the user should initiate the remedy. The problem may also be alleviated by switching to higher order elements, such as quadratic or cubic elements.

Example applications: Figure 4.17 shows typical examples of applications for shell problems.

REMARKS

- Shell structures are general 3-D surfaces of small thickness. Loading can be in plane and out of plane. All stress components except those normal to the surface will be nonzero.
- The shell element may be triangular with 3, 6, or 9 nodes or quadrilateral with 4, 8, or 12 nodes.
- Commercial programs offer shell elements with variable thickness and with membrane and/or bending capabilities.
- Linear elements may experience "shear locking": nonphysical, high stiffness in shear. Various remedies are available in most programs.

4.1.6   Three-Dimensional Solid Problems

By default, if the problem is not one of those discussed in Section 4.1.3 to Section 4.1.5 then it will be classified as a 3-D problem. Three-dimensional problems are easily identified by their 3-D geometry and loading, as shown in Figure 4.18. Any of the categories of problems discussed above may be solved by using 3-D elements. The critical drawback is the substantial increase in analyst time for modeling and in CPU time in processing the solution (the time needed may be one order of magnitude larger than for a corresponding 2-D problem).

The following summarizes the conditions for a 3-D problem.

- *Geometry*:
  - A 3-D object with no apparant thickness or uniformity in any direction.
  - The boundary of the object may be straight or curved.

Thin-walled cylinder with 3D loading

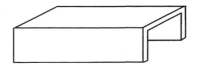

C-Channel with 3D loading

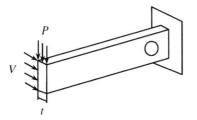

Cantliver under in-plane and out-of-plane loading

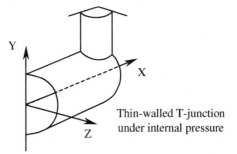

Thin-walled T-junction under internal pressure

FIGURE 4.17 Typical shell and plate examples.

- *Loading*:
  - Three-dimensional loading. An important note should be added here: 3-D elements normally have three displacement DoF and no rotational DoF. This means that moments cannot be directly applied to such elements and the moment effect should be simulated via concentrated forces or couples.
  - Other loading, including thermal loading and body or inertia forces, may be applied in any direction.

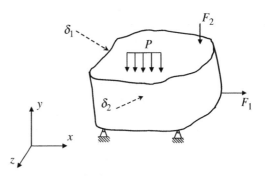

FIGURE 4.18 Three-dimensional problems.

Degrees of freedom: Three-dimensional solid elements have three translational DoF: u_x, u_y, and u_z. As mentioned above, concentrated moments should be modeled by an equivalent system of concentrated forces or couples.

Element shapes: Figure 4.19 shows some typical 3-D element shapes available in most commercial FE programs.

Example applications: Figure 4.20 shows some typical examples of 3-D structures requiring 3-D solid modeling and elements. It should be noted that if out-of-plane vibration modes of a 2-D structure are of concern, then a 3-D solid or shell model should be used even if the loading is in one plane.

Remarks

- Structures that are not classified as truss, beam, 2-D, or shell may be modeled with 3-D solid elements.
- Three-dimensional elements have three translational DoF per node (u_x, u_y, and u_z) and may have 4–32 nodes per element.

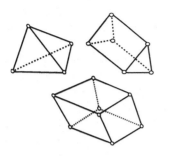
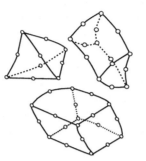
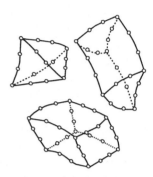
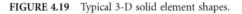

4-node tetrahedron,	10-node tetrahedron	16-node tetrahedron
6-node wedge and	15-node wedge and	21-node wedge and
8-node cube (linear)	20-node cube (quadratic)	32-node cube (cubic)

FIGURE 4.19 Typical 3-D solid element shapes.

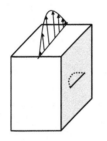

Shaft with torque, bending
moment and shear forces

Welded joint with three
dimensional loading

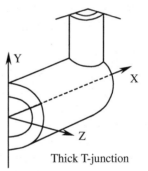

Thumb nail crack in a thick block

Thick T-junction

FIGURE 4.20 Typical 3-D examples.

4.1.7   Synopsis of Problem Classification and Element Choice

Table 4.1 summarizes the concepts discussed above (Section 4.1.2 to Section 4.1.6) for element choice and problem classification. The table also shows displacement and stress variations within each element as well as standard element output quantities. The displacement variation within the element represents the basic element assumption in FE analysis and is called the "shape function" or the "approximation function" assumption. The strain variation within the element follows by differentiating the displacement variation according to the strain–displacement relations. The number of nodes per element is linked to the shape function assumption. For example, in 2-D elements three-node triangles have linear shape functions and in 3-D elements eight-node bricks have linear shape functions. Elements with linear shape function are sometimes called low-order or linear elements. Such elements may be used quite efficiently in both linear and nonlinear analyses.

TABLE 4.1 Summary of Problem Classification and Corresponding Element Characteristics

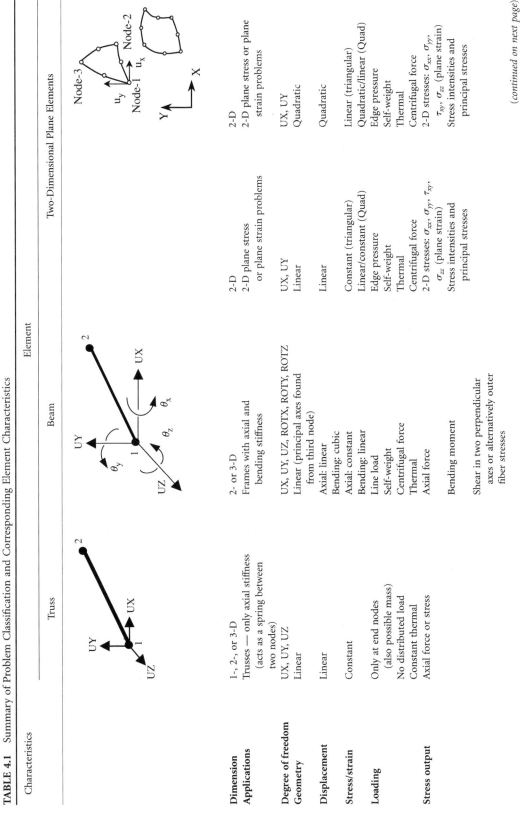

Characteristics	Truss	Beam	Two-Dimensional Plane Elements	
Dimension	1-, 2-, or 3-D	2- or 3-D	2-D	2-D
Applications	Trusses — only axial stiffness (acts as a spring between two nodes)	Frames with axial and bending stiffness	2-D plane stress or plane strain problems	2-D plane stress or plane strain problems
Degree of freedom Geometry	UX, UY, UZ Linear	UX, UY, UZ, ROTX, ROTY, ROTZ Linear (principal axes found from third node)	UX, UY Linear	UX, UY Quadratic
Displacement	Linear	Axial: linear Bending: cubic	Linear	Quadratic
Stress/strain	Constant	Axial: constant Bending: linear	Constant (triangular) Linear/constant (Quad)	Linear (triangular) Quadratic/linear (Quad)
Loading	Only at end nodes (also possible mass) No distributed load Constant thermal	Line load Self-weight Centrifugal force Thermal	Edge pressure Self-weight Thermal Centrifugal force	Edge pressure Self-weight Thermal Centrifugal force
Stress output	Axial force or stress	Axial force Bending moment Shear in two perpendicular axes or alternatively outer fiber stresses	2-D stresses: σ_{xx}, σ_{yy}, τ_{xy}, σ_{zz} (plane strain) Stress intensities and principal stresses	2-D stresses: σ_{xx}, σ_{yy}, τ_{xy}, σ_{zz} (plane strain) Stress intensities and principal stresses

(continued on next page)

TABLE 4.1 (*continued*)

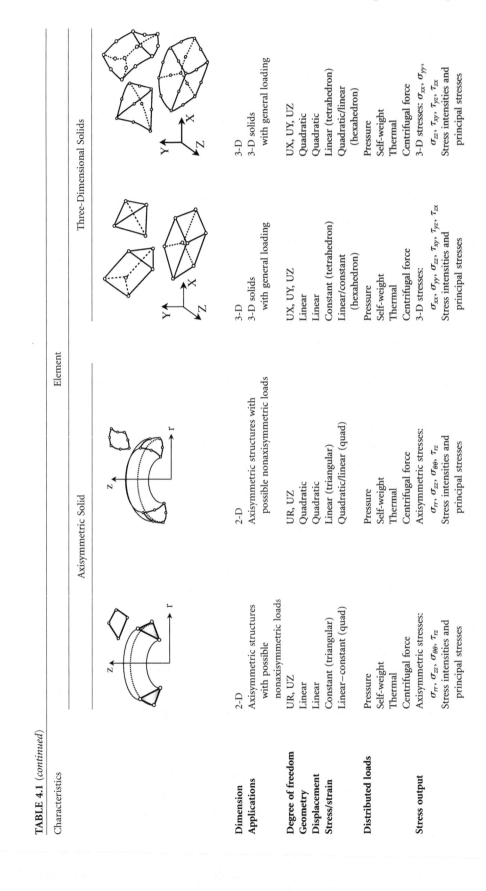

Characteristics	Element			
	Axisymmetric Solid		Three-Dimensional Solids	
Dimension	2-D	2-D	3-D	3-D
Applications	Axisymmetric structures with possible nonaxisymmetric loads	Axisymmetric structures with possible nonaxisymmetric loads	3-D solids with general loading	3-D solids with general loading
Degree of freedom	UR, UZ	UR, UZ	UX, UY, UZ	UX, UY, UZ
Geometry	Linear	Quadratic	Linear	Quadratic
Displacement	Linear	Quadratic	Linear	Quadratic
Stress/strain	Constant (triangular) Linear–constant (quad)	Linear (triangular) Quadratic/linear (quad)	Constant (tetrahedron) Linear/constant (hexahedron)	Linear (tetrahedron) Quadratic/linear (hexahedron)
Distributed loads	Pressure Self-weight Thermal Centrifugal force	Pressure Self-weight Thermal Centrifugal force	Pressure Self-weight Thermal Centrifugal force	Pressure Self-weight Thermal Centrifugal force
Stress output	Axisymmetric stresses: σ_{rr}, σ_{zz}, $\sigma_{\theta\theta}$, τ_{rz} Stress intensities and principal stresses	Axisymmetric stresses: σ_{rr}, σ_{zz}, $\sigma_{\theta\theta}$, τ_{rz} Stress intensities and principal stresses	3-D stresses: σ_{xx}, σ_{yy}, σ_{zz}, τ_{xy}, τ_{yz}, τ_{zx} Stress intensities and principal stresses	3-D stresses: σ_{xx}, σ_{yy}, σ_{zz}, τ_{xy}, τ_{yz}, τ_{zx} Stress intensities and principal stresses

TABLE 4.1 (*continued*)

Characteristics	Element	
	Three-Dimensional Shell Structures	

Dimension	3-D	3-D
Applications	3-D shell structures with general loading	3-D shell structures with general loading
Degree of freedom	UX, UY, UZ, ROTX, ROTY, ROTZ	UX, UY, UZ, ROTX, ROTY, ROTZ
Geometry	Linear	Quadratic
Displacement	In shell local coordinates: linear on midsurface and linear through the thickness	In shell local coordinates: quadratic on midsurface and linear through the thickness
Stress/strain	On the midsurface and in shell local coordinates: Triangle: constant Quad: linear/constant (linear through thickness)	On the midsurface and in shell local coordinates: Triangle: linear Quad: quadratic/linear (linear through thickness)
Distributed loads	Surface and edge pressure Self-weight Thermal Centrifugal force	Surface and edge pressure Self-weight Thermal Centrifugal force
Stress output	3-D stresses σ_{xx}, σ_{yy}, σ_{zz}, τ_{xy}, τ_{yz}, τ_{zx} (global) Local stresses Stress intensities and principal stresses	3-D stresses σ_{xx}, σ_{yy}, σ_{zz}, τ_{xy}, τ_{yz}, τ_{zx} (global) Local stresses Stress intensities and principal stresses

4.2   Types of Analysis

If the applied loading on the structure is to change with time, the designer should make a decision on whether or not a dynamic analysis is required. To be able to make this decision, information about the loading and the natural frequencies of the structure are required. From the loading point of view, we classify a general loading on a structure into one of four categories: steady, cyclic, transient, and random.

Figure 4.21 shows a schematic presentation of the four loading categories. Figure 4.21b shows two types of cyclic loading, one is a simple harmonic loading with amplitude of oscillation F_o, period $T = 2\pi$, and frequency $f = 1/T$. The second is a periodic loading with a period T. Using Fourier analysis, any periodic function with a period, T, may be decomposed into a series of harmonic sines and cosines with frequencies $f_1 = 1/T, 2f_1, 3f_1, \ldots, nf_1$. Figure 4.21c shows a transient load with a duration T.

Some forcing functions do not lend themselves to simple frequency or time specifications. Figure 4.21d shows a typical time history of a forcing function of such a category that may be considered as random excitation. Transient and cyclic forcing functions are normally specified as force vs. time or frequency, respectively. On the other hand, random forcing functions are commonly specified as the magnitude of the input acceleration squared vs. frequency. This input data is normally called a power spectral density (PSD) input curve. Time history input may be used in the analysis of random excitation and the random input may be treated as transient. This would normally require extensive computer resources and CPU time due to the very large number of time steps that would be required to capture the peak response. Random excitation may occur from random sources such as road undulation on vehicles, noise, earthquakes, and seismic events on buildings, and wind and turbulent loading on airplanes. For practical purposes, PSD curves have been compiled for various random events and are normally available in most FE commercial, programs.

In addition to the load specification, the second factor affecting the decision on the type of analysis is the natural frequency of the structure. Structures with mass and stiffness characteristics are capable of free vibrations after removing the initial excitation on the structure. Depending on the initial conditions of excitation, the structure may vibrate in one natural frequency or in a combination of more than one

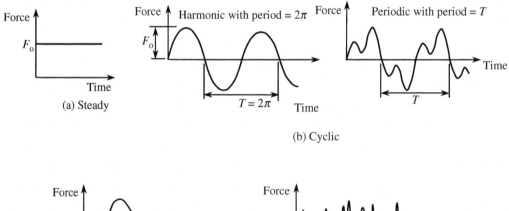

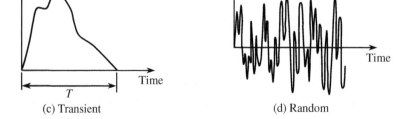

FIGURE 4.21   Various loading types.

natural frequency. In the latter case, the vibration will result in a complex periodic motion that may be decomposed into a series of harmonic motions vibrating at the first natural frequency (fundamental one) and its multiples. As indicated, these frequencies are called the natural or resonant frequencies of the system and they depend on the mass and elastic or stiffness characteristics of the system. In other words, the natural frequencies of a system are those frequencies that the system tends to vibrate under conditions of free vibration. Theoretically, a continuous system or structure has infinite DoF and infinite natural frequencies. From a practical modeling point of view, the structure will have a number of natural frequencies equal to the number of DoF used to model the structure. The mode shapes give the relative displacements of each point in the structure when it vibrates in one of the natural frequencies. Each natural frequency has a corresponding mode shape. The first mode shape corresponding to the first natural frequency of a structure represents the most flexible way in which the structure may deform or vibrate and corresponds to the least strain energy level in all modes. It is important to note that, a symmetric structure will have symmetric and antisymmetric mode shapes. This may be realized by considering the mode shapes of a simple beam as shown in Figure 4.22 where all odd number modes are symmetric and all even number modes are antisymmetric.

The information provided by the natural frequencies and mode shapes of a structure is vital in understanding the behavior of the structure under general excitation. If a structure is excited in one of its natural frequencies, then theoretical analysis of the structure response as an undamped system shows that the amplitude of the resulting vibration response reaches infinity. In practice, all structures possess a certain amount of damping that will limit the amplitude of vibration.

Determining the type of analysis required for a structure depends on the nature of the applied load and the magnitude of the first natural frequency or the fundamental frequency of the structure. The two main categories of analysis types are static and dynamic analyses. A steady load, i.e., a load that does not change with time (Figure 4.21a), would require simple static analysis. If the load varies with time, it does not necessarily mean that dynamic analysis needs to be performed. For example, if the loading is harmonic with a frequency less than approximately one third of the first natural frequency of the structure then static analysis will provide an accurate solution and we only need to solve the static problem for the peak load values. (For this frequency range, static analysis may provide a maximum difference of 12.5% in the response for undamped structures and for a forcing frequency less than one fourth of the natural frequency, the difference is only 6.7%.) In this case, the load is normally called "quasi-static." We may apply the same rule to periodic loading after decomposing it into its harmonic components. In transient loading, we should consider the time of application of the load or the "rise time." If the longest natural period corresponding to the first natural frequency of the structure is more than about twice the rise time, the loading should be classified as shock or impact loading and transient dynamic analysis would be required. If the longest natural period of a system is less than about one third of the rise time, it would be sufficient to perform static analysis and consider the loading to be quasi-static. If the longest natural period falls between the quasi-static and shock conditions then a transient dynamic analysis would be required and the load would be classified as transient loading. If the loading cannot be categorized as frequency dependent or time dependent, as for example the one shown schematically in Figure 4.21d, then it should be considered random.

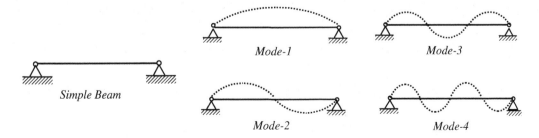

FIGURE 4.22   First four mode shapes of a simple beam.

From the above discussion, the following categories of dynamic analysis may be classified (Mirovitch, 1980):

Modal and natural frequency analysis: This is normally performed before any other type of dynamic analysis and will produce the natural frequencies and mode shapes of the structure. This information is vital for understanding the dynamic behavior of the structure and provides data for decoupling the dynamic equations in other analyses. One objective of this analysis is to make sure that the structure is not operating at a frequency close to one of its natural frequencies. A comfortable range for an operating frequency is three times higher than the nearest natural frequency from the lower side and three times smaller than the nearest natural frequency from the higher side. If the operating frequency is higher than the first natural frequency, then large vibration amplitudes or "shudder" will be evident upon passing the natural frequency and certain startup procedures may have to be devised.

Frequency response analysis: This type of analysis is performed if the loading on the structure is harmonic or periodic. In periodic loading, the loading function is first decomposed into its sine and cosine components using Fourier analysis. In this analysis, the response will be harmonic with the same loading frequency and with a possible phase shift. The output of such analysis would be displacements, velocities, and accelerations that may be used to calculate forces and stresses in the structure. The displacement output of this analysis defines the deformed shape of the structure, which, in general, is different from the structural mode shapes and may be used to calculate stresses and strains in the structure. Assuming linear conditions, the response to multiple frequency inputs may be simply summed up using the superposition technique.

Transient response analysis: This type of analysis is performed if the loading on the structure is classified as transient or shock. There are two general approaches to solving the equation of motion in transient analysis. One is the "direct integration approach" and the other is the "modal superposition approach." In the direct integration approach, the system equations of motion are integrated directly in the time domain. The required number of time steps depends on the period and the assumptions specified for displacement, velocities, and accelerations within the time step. This number may be quite large and the solution to large size problems may become a computationally difficult task.

In the second method, the modal superposition approach, the equations of motion are first transformed to modal generalized displacements. In order to perform this transformation, the mode shapes of the structure should first be determined through a modal analysis. The transformation yields a set of decoupled second-order differential equations for the system that are easier to solve. The basic assumption in this approach is that the superposition of the structural response due to the first few lower mode shapes adequately represents the dynamic response of the structure under general transient loading. In practice, this means that only a fraction of the mode shapes of the structure are needed to accurately represent the dynamic behavior. This approach will be generally less accurate than the direct integration approach but will provide substantial savings in computation time for large size problems.

The output of transient analysis is time histories of displacements, velocities, and accelerations of the system that may be used to calculate forces and stresses.

Random response analysis: If the loading on the structure cannot be classified as frequency or time dependent, it is considered random. Problems with random loading may be solved using a "time history approach" or a "power spectral density (PSD) approach." The time history approach treats the random input as a transient one and performs a step-by-step integration over the excitation period. This is normally very costly and requires intensive use of computational resources. If the structure has multiple random inputs in more than one direction, then the transient time history approach is more accurate and may be the preferred approach to solve the problem. Structures with random inputs in more than one direction may be analyzed by a modal superposition approach with special procedures to combine the modal responses (see Section 4.5.4 for details). In the PSD approach, the input will be acceleration as a function of frequency. This is normally specified as a discrete or continuous spectrum by providing values of the (G^2/f) vs. f, where G is the input acceleration and f is the frequency. This approach is most appropriate for a single input in one direction.

As mentioned above, most programs will have standard time history inputs or PSD inputs to simulate commonly used random excitations such as vibrations due to road irregularities, earthquakes and seismic inputs, vibration due to wind loads for aerospace parts, and wind loads for wind tunnel tests.

REMARKS

- Structural loading may be classified into four categories: steady (constant with time), cyclic (harmonic with period $T = 2\pi$ and frequency $f = 1/T$ or general periodic that may be decomposed into harmonics by Fourier analysis), transient with a duration T, and random.
- Random forcing functions are commonly specified as the magnitude of the input acceleration squared vs. frequency (PSD). Typical PSD curves have been compiled for various random events and are normally available in commercial FE programs.
- The natural frequencies of a system are those frequencies that the system tends to vibrate under conditions of free vibration. The mode shapes give the relative displacements of each point in the structure when it vibrates in one of the natural frequencies. Symmetric structures will generally have symmetric and antisymmetric mode shapes.
- If the loading is harmonic with a frequency of less than approximately one third of the first natural frequency of the structure, static analysis for the peak load levels suffices. In transient loading, if the longest natural period of a system is less than about one third of the rise time, loading is quasi-static and static analysis suffices.
- In transient loading, if the longest natural period corresponding to the first natural frequency of the structure is more than about twice the rise time of the load, it is necessary to classify the loading as shock or impact loading and transient dynamic analysis would be required.
- If the longest natural period falls between the quasi-static and shock conditions, a transient dynamic analysis would be required.

4.3 Modeling Aspects for Dynamic Analysis

Section 4.1 and Section 4.2 provide guidelines for problem and analysis classification. Once a decision has been made about the type of problem and the method of analysis to be used, the designer still has to finalize many details to create a working model for the structure. This section provides a brief discussion of the main aspects required to do this. This includes choice of model size and master DoF, lumped and consistent mass modeling, and use of symmetry. Another important aspect of modeling, damping, will be dealt with in Section 4.4.

4.3.1 Model Size and Choice of Master Degrees of Freedom

Using the same model for both static and dynamic analysis will make the FE solution to the dynamic problem more complicated and therefore much more memory and CPU resources will be required. Also, the results may not be necessarily more useful or accurate. Therefore, FE models for dynamic analysis should, in general, be kept simple.

Depending on the type of dynamic analysis, it may be possible to construct a much simpler model for the dynamic problem, possibly using spring, beam, and mass elements. If we consider modal analysis, for example, the objective is to find the mode shapes of the structure and the corresponding natural frequencies. In practice, we seldom need more than the very few lower natural frequencies and mode

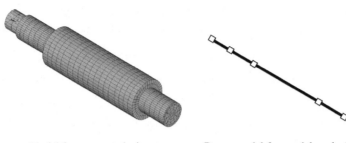

Model for stress analysis   Beam model for modal analysis

FIGURE 4.23   Static and dynamic models for a shaft.

shapes for a structure. Using a static model that may have thousands of DoF in a modal analysis is very time consuming and does not serve the purpose of the mode shape analysis. It may be better and more efficient to construct a mass spring/beam model for the modal analysis. Figure 4.23 shows a static analysis model for a shaft and a corresponding much simpler mode shape analysis model. In the mass/beam model for modal analysis, the beam properties may change to accommodate specific variations in the shaft geometry. It is also possible to extract accurate values for the spring and beam stiffnesses in 3-D using a detailed static model.

Most programs provide means for condensing or reducing the DoF in a dynamic analysis. A common condensation method used is the "Guyan reduction" (see e.g., Freed and Flanigan, 1991; Bouhaddi and Fillod, 1992; Deiters and Smith, 2000). In these condensation methods, the analyst specifies certain DoF to be masters and to be retained by the program, while all other DoF would be considered slaves and would be eliminated or condensed. The dynamic analysis is then carried out using the master DoF. Information pertaining to slave DoF is still available and may be extracted in subsequent runs that may involve stress or other detailed analyses or postprocessing. Reduction has the same effect as modal superposition in terms of reducing the problem size and the required CPU time.

Fundamental differences between the two methods may be realized due to the nature of reducing the system of equations. In the modal superposition technique, reduction is accomplished by assuming that the superposition of the structural response due to the first few lower mode shapes adequately represents the response of the structure. In other reduction methods, such as the Guyan reduction technique, we assume that the response of the system is accomplished from the response of certain predetermined master DoF. In the modal superposition technique, a modal analysis must be performed first and the final equations of motion are decoupled, whereas in the other reduction methods no modal analysis is necessary and the final equations of motion are generally full and coupled. Reduction of the total DoF to a set of master DoF has the effect of imposing displacement constraints on the system that increase the stiffness and subsequently overestimate the natural frequencies. The least affected modes are the lower ones. Therefore, reduction and modal superposition techniques should not be used, in general, for shock and impact loading (the modal superposition technique entirely ignores the response from higher modes).

The choice of master DoF may be performed automatically by the program. There is no assurance, however, that this will always produce the best choice of masters. The analyst may like to use the program for a first choice of master DoF and then augment the choice manually to improve the accuracy. Simple rules may be used to identify the master DoF (see *ANSYS User's Manual*, 2003). The general rule is to choose DoF with large mass and small stiffness or large mass to stiffness ratios as masters. Master DoF should be chosen in the direction of the expected response of the structure, e.g., in a beam or shell model, lateral DoF would be more appropriate than axial or membrane DoF, assuming that the load produces lateral deflections. Rotational DoF are seldom used as masters. It is also important to spread out and not cluster the master DoF so that they will be capable of producing the structural response. Finally, DoF with concentrated mass, a specified input force, or displacement should be retained as masters.

The question of how many master DoF are adequate is not an easy one to answer. The rule of thumb is to have having at least double the number of master DoF compared with the number of modes of interest. The other guideline is to use 10 to 50% of the total DoF as masters. In practice, the only sure way to test the adequacy of the number of master DoF is to run the analysis with two different choices and compare the results to make sure that the modes of interest are not substantially different.

REMARKS

- In general, dynamic analysis models should be simpler and much smaller in terms of the number of DoF (than static analysis ones). In modal analysis, it is better and more efficient to construct a mass spring/beam model for the structure.
- "Guyan reduction" may be used to reduce the model size by selecting master DoF for the analysis and condensing out slave DoF.
- Master DoF may initially be chosen by the program. The analyst should, however, ensure that the choice includes nodes with a large mass to stiffness ratio and with DoF in the direction of the expected response of the structure. Also, it is important to spread out and not cluster the master DoF.
- As a rule of thumb, the number of master DoF should be at least double the number of modes of interest or about 10 to 50% of the total DoF.

4.3.2 Lumped and Consistent Mass Modeling

The mass characteristic of the structure may be modeled as a lumped or consistent mass matrix. Lumped mass simply means dividing the total mass of the element by the number of nodes. This produces a diagonal mass matrix and may significantly reduce the numerical effort required to solve the dynamic equations. Considering rotational inertia terms in lumped mass assumptions is arbitrary and is normally ignored. If the lumped mass matrix has zero diagonal terms, the matrix will be positive semidefinite and the natural periods corresponding to the zero diagonal terms will be zero. This will have serious implications on the choice of the time step in explicit direct integration solution methods and will be discussed in Section 4.4. A lumped mass assumption implies a discontinuous acceleration field in the element and it usually gives good accuracy for lower modes and frequencies (Chan et al., 1993; Jensen, 1996). A consistent mass matrix is derived by introducing inertia forces in the virtual work formulation of a dynamic FE problem. This would normally lead to a full mass matrix making contributions to both translational and rotational DoF as well as coupling terms. Consistent mass representation usually provides better accuracy for higher modes and frequencies (Kim, 1993).

It may be concluded that, if no reduced integration is used to compute element stiffness, the element is compatible, and a consistent mass matrix is used, then the calculated natural frequencies will be an upper bound of the model frequencies (Cook et al., 1989). In some applications such as plates, shells, and 3-D solid elements, it has been found that the convergence rate of the natural frequency calculations for the lumped mass discretization is the same as that for a consistent mass formulation (Chan et al., 1993; Jensen, 1996).

Some FE programs provide a modified lumped mass modeling in which the diagonal translational terms are proportional to the diagonal terms of the corresponding consistent mass matrix. This yields more accurate results, in general, than the traditional lumped mass assumption. Several studies have been published on the merits of combining consistent and lumped mass formulations into some kind of modified lumped mass matrix (Chan et al., 1993; Kim, 1993; Jensen, 1996). The modified lumped mass matrix is generally a linear combination of the lumped and the consistent mass matrices. For an improved accuracy, the consistent mass should be weighted more than the lumped mass.

In building a lumped mass model for the structure, the analyst should realize that one of the objectives is to capture the few lower natural frequencies and mode shapes of the structure. To achieve this, the model should minimize the strain energy and maximize the kinetic energy of the structure. Minimizing the strain energy of the structure may be realized by accurately modeling the soft links in the load path. Structural supports are particularly difficult in this regard. Changing the support model from completely rigid to slightly flexible may shift the first natural frequency by as much as 40% (although it will not have much impact on the mode shape). Maximizing the kinetic energy is simply realized by accurate modeling of the system mass, especially in areas where the mass is large and the stiffness is small.

Care should be exercised when modeling masses attached to the system, e.g., a large "nonstructural" mass on top of a building (e.g., a mass of a machine, reservoir, etc.) or the mass of an engine in a car model. If the stiffness characteristic of the nonstructural mass is not important, it may be simply modeled as a rigid lumped mass with an offset mass center from the nearest node. Most programs have a rigid link capability to connect this mass to the structure. This kind of model is transformed internally to a nondiagonal mass matrix and may only be used with a consistent mass formulation for the structural mass. If the structural mass is modeled as lumped, the nonstructural mass will simply be added to the nearest node, which may not be an accurate representation of the physical model.

REMARKS

- A lumped mass matrix is a diagonal mass matrix that may be obtained by simply dividing the total mass by the number of DoF or alternatively by certain combinations of lumped and consistent mass terms. A consistent mass matrix is derived from FE virtual work equations and is normally a full mass matrix having translational, rotational, and coupling terms.
- Lumped mass may significantly reduce the numerical effort required to solve the dynamic equations without impacting the accuracy.
- Zero diagonal terms in the mass matrix should be avoided as much as possible.
- A lumped mass assumption usually gives good accuracy for lower modes and frequencies, whereas consistent mass representation usually gives better accuracy for higher modes and frequencies.

4.3.3 Use of Symmetry

Use of symmetry to reduce the analysis cost and computer time is rather tricky in modal analysis. As shown in Table 4.2, symmetric structures have both symmetric and nonsymmetrical mode shapes. If we only use a symmetric model with symmetric boundary conditions, all nonsymmetrical modes will be undetected. This will change the interpretation of the results and will have an impact on subsequent analyses using the modal data. If the objective is to perform only a modal analysis, it may be advisable to have a less detailed full model rather than a detailed half or quarter model. Also, if the

TABLE 4.2 Symmetric and Antisymmetric Boundary Conditions

Plane of symmetry	Symmetric Boundary Conditions						Antisymmetric Boundary Conditions					
	u_x	u_y	u_z	θ_x	θ_y	θ_z	u_x	u_y	u_z	θ_x	θ_y	θ_z
XY	—	—	0	0	0	—	0	0	—	—	—	0
XZ	—	0	—	0	—	0	0	—	0	—	0	—
YZ	0	—	—	—	0	0	—	0	0	0	—	—

objective is to perform frequency response or transient analysis using the modal data, it may be advisable to rely on a full model. Experienced analysts may use symmetric models and run the modal analysis with symmetric and antisymmetric boundary conditions to capture the response of all modes. Table 4.2 summarizes symmetric and antisymmetric boundary conditions for translational and rotational DoF.

4.4   Equations of Motion and Solution Methods

In this section, an overview of the equation of motion and the various solution methods available are provided. In most cases, detailed derivation of the equations is not given. The intention is to highlight various solution schemes and the practical aspects of choosing a particular one for a given system or a given dynamic analysis. A consideration of damping and its idealization are also discussed in this section.

4.4.1   Equation of Motion

The equations of motion describing the dynamic behavior of a structural system may be obtained from the extended Hamilton's principle for elastodynamics:

$$\delta \int_{t_o}^{t_f} (U - T)\mathrm{d}t - \int_{t_o}^{t_f} \delta W \, \mathrm{d}t = 0 \tag{4.1}$$

where

δ: first variation
t_o, t_f: two arbitrary time points at which the first variation vanishes
U: strain energy of the system
T: kinetic energy of the system
δW: virtual work of the external forces acting on the system during virtual displacements

The external forces may be conservative or nonconservative. Nonconcentrative forces, e.g., damping and follower forces, are deformation dependent, i.e., their magnitudes and/or directions depend on the deformations. Introducing the finite element discretization assumption and substituting expressions for the virtual work and the kinetic and strain energy terms into the above equation, we obtain the Lagrange equation which finally leads to the equations of motion of the system that may be written in the following form:

$$\mathbf{M\ddot{u}} + \mathbf{C\dot{u}} + \mathbf{Ku} = \mathbf{p(t)} \tag{4.2}$$

where $\mathbf{M}$, $\mathbf{C}$, and $\mathbf{K}$ are, respectively, the global mass, global damping, and global stiffness matrices (made up by proper assembly of the element matrices); $\mathbf{p}(t)$ is the time-dependent applied force vector; and $\mathbf{\ddot{u}}$, $\mathbf{\dot{u}}$, and $\mathbf{u}$ are the nodal acceleration, nodal velocity, and nodal displacement vectors, respectively.

The global mass and damping matrices are assembled from the element matrices that are given by (Cook et al., 1989; Bathe, 1996):

$$\mathbf{M}^{(e)} = \int_V \rho^{(e)} \mathbf{N}^{\mathrm{T}} \mathbf{N} \, \mathrm{d}V \tag{4.3}$$

$$\mathbf{C}^{(e)} = \int_V c_s^{(e)} \mathbf{N}^{\mathrm{T}} \mathbf{N} \, \mathrm{d}V \tag{4.4}$$

where $\rho^{(e)}$ and $c_s^{(e)}$ are the mass density and the viscous damping coefficient for an element e, and $\mathbf{N}$ is the matrix of the element shape functions.

As indicated above, there are two general approaches to solving Equation 4.1. One is the direct integration approach and the other is the modal superposition approach. With the direct integration approach, the equations of motion are integrated directly in the time domain. In the second approach, the modal superposition method, the equations of motion are first transformed to modal generalized displacements through the use of the mode shapes of the structure. The transformation yields a set of decoupled second-order differential equations for the system that are easier to solve. In the remaining part of this section, we describe the two solution approaches in more detail.

REMARKS

- Equation 4.2 is the general equation of motion for a dynamic system. Two solution methods are available: the direct integration approach and the modal superposition approach.

4.4.2 Direct Integration Method

With this method, the coupled equations of motion are integrated directly without special transformation. Direct integration schemes seek to satisfy the equilibrium equations of motion at discrete time points, $0, \Delta t, 2\Delta t, \ldots, t, t + \Delta t, \ldots$ etc. This requires assumptions for the variation of $\ddot{\mathbf{u}}$, $\dot{\mathbf{u}}$, and $\mathbf{u}$ within the time step. These assumptions will determine the accuracy and convergence characteristics of a particular time integration scheme. They also govern the variation of these quantities with time and are different from the shape function assumptions for the element, which govern the displacement variation within the element, i.e., with spatial coordinates. Direct integration schemes may be categorized as "explicit" and "implicit" schemes. In explicit schemes, the equilibrium equations at time t are used to solve the unknowns at time $t + \Delta t$. These schemes are usually fast and efficient and require no factorization of the effective stiffness matrix provided that the mass and damping matrices can be formulated as diagonal matrices. The main drawback of these schemes is the requirement that the time step for the analysis has to be smaller than a critical value. On the other hand, in implicit schemes, the equilibrium equations at time $t + \Delta t$ are used to solve unknowns at time $t + \Delta t$. These schemes require factorization of the effective stiffness matrix of the structure at each time step but do not require the condition of a critical time step. In implicit schemes, the number of operations at each time step is proportional to the matrix order times half of the band width, $m = \alpha nb$, where m is the number of operations, n is the number of equations, α is a constant ≥ 2, and b is half the band width.

Many integration schemes are available in commercial FE programs. These include Euler forward, Euler backward, central difference, Houbolt, Wilson-θ, and Newmark schemes (Belytschko and Hughes, 1983; Zeng et al., 1992). In the following, one example of an explicit scheme and one example of an implicit scheme are discussed.

4.4.2.1 The Central Difference Method

This is an explicit scheme in which the velocity and acceleration assumptions are given by

$$\left.\begin{array}{l} {}^{t}\dot{\mathbf{u}} = \dfrac{1}{2\Delta t}\left({}^{t+\Delta t}\mathbf{u} - {}^{t-\Delta t}\mathbf{u}\right) \\[2mm] {}^{t}\ddot{\mathbf{u}} = \dfrac{1}{(\Delta t)^2}\left({}^{t+\Delta t}\mathbf{u} - 2\,{}^{t}\mathbf{u} + {}^{t-\Delta t}\mathbf{u}\right) \end{array}\right\} \tag{4.5}$$

Using Equation 4.5 in the equilibrium equation at time (t) to obtain a solution at time $(t + \Delta t)$ gives

$$\left[\frac{1}{(\Delta t)^2}\mathbf{M} + \frac{1}{2(\Delta t)}\mathbf{C}\right]{}^{t+\Delta t}\mathbf{u} = {}^{t}\mathbf{p} - \left[\mathbf{K} - \frac{2}{(\Delta t)^2}\mathbf{M}\right]{}^{t}\mathbf{u} - \left[\frac{1}{(\Delta t)^2}\mathbf{M} - \frac{1}{2(\Delta t)}\mathbf{C}\right]{}^{t-\Delta t}\mathbf{u} \tag{4.6}$$

If the mass and damping matrices are diagonal, Equation 4.6 requires no assembly and no factorization of the effective stiffness matrix of the structure is required. The scheme is, however, conditionally stable and requires a time step smaller than a critical value given by

$$\Delta t_{\mathrm{cr}} \leq T_n/\pi \tag{4.7}$$

where T_n is the smallest period in the FE assembly with n DoF. This is an important condition and should be examined carefully. If the mass matrix has a zero diagonal value, the corresponding period will be zero and Equation 4.7 cannot be satisfied. Introducing very small diagonal values may not provide a practical solution since the time step will be very small and the number of steps will be excessive. The scheme is generally applied when a lumped mass assumption with no zero diagonal terms and velocity-dependent damping are assumed. The initial conditions required for this scheme are the displacements, velocities, and accelerations at time $t = 0$.

4.4.2.2 The Newmark Method

This is a commonly used implicit scheme with the following assumptions:

$$^{t+\Delta t}\dot{\mathbf{u}} = {}^{t}\dot{\mathbf{u}} + \left[(1 - \alpha){}^{t}\ddot{\mathbf{u}} + \alpha {}^{t+\Delta t}\ddot{\mathbf{u}}\right]\Delta t \tag{4.8}$$

$$^{t+\Delta t}\mathbf{u} = {}^{t}\mathbf{u} + {}^{t}\dot{\mathbf{u}}\,\Delta t + \left[\left(\frac{1}{2} - \beta\right){}^{t}\ddot{\mathbf{u}} + \beta {}^{t+\Delta t}\ddot{\mathbf{u}}\right](\Delta t)^2 \tag{4.9}$$

where α and β are user input parameters. Equation 9.8 and Equation 4.9 are solved for $^{t+\Delta t}\dot{\mathbf{u}}$ and $^{t+\Delta t}\ddot{\mathbf{u}}$, then substituted into the equilibrium equations at time $(t + \Delta t)$ to obtain:

$$\left[\mathbf{K} + \frac{\alpha}{\beta\Delta t}\mathbf{C} + \frac{1}{\beta(\Delta t)^2}\mathbf{M}\right]{}^{t+\Delta t}\mathbf{u} = {}^{t+\Delta t}\bar{\mathbf{p}} \tag{4.10}$$

where the right-hand side is a function of the parameters α, β, Δt, $\mathbf{K}$, $\mathbf{M}$, $\mathbf{C}$, ${}^{t}\mathbf{u}$, ${}^{t}\dot{\mathbf{u}}$, and ${}^{t}\ddot{\mathbf{u}}$. Equation 4.10 indicates that the scheme requires assembly and factorization of the effective stiffness matrix of the structure. The choice of parameters α and β will determine the stability of the scheme and may reduce the scheme to be equivalent to other known ones as follows:

- For $2\beta \geq \alpha \geq 1/2$, the scheme is unconditionally stable. This does not guarantee accuracy and only ensures that the results will not grow out of bounds (Belytschko and Hughes, 1983).
- A slightly more stringent criterion for unconditional stability that provides artificial damping in higher modes is given by using (Belytschko and Hughes, 1983):

$$\alpha \geq \frac{1}{2} \quad \text{and} \quad \beta \geq \frac{1}{4}\left(\alpha + \frac{1}{2}\right)^2$$

- A commonly used option gives $\alpha = 1/2$ and $\beta = 1/4$ which corresponds to a trapezoidal rule with constant average acceleration.
- Another option that corresponds to a linear acceleration assumption is given by $\alpha = 1/6$ and $\beta = 1/2$. This method is very good for small Dt but tends to be unstable for large Dt.

Table 4.3 summarizes the above choices for α and β.

TABLE 4.3 Typical Choices of α and β Parameters for the Newmark Scheme

α Value	β Value	Comments
$2\beta \geq \alpha \geq 1/2$	$2\beta \geq \alpha \geq 1/2$	No guarantee of accuracy
$\alpha \geq \dfrac{1}{2}$	$\beta \geq \dfrac{1}{4}(\alpha + \dfrac{1}{2})^2$	Improved accuracy Provides artificial damping in higher modes
$\alpha \geq \dfrac{1}{2}$	$\beta \geq \dfrac{1}{4}(\alpha + \dfrac{1}{2})^2$	Improved accuracy Corresponds to a trapezoidal rule with constant average acceleration
$\alpha \geq \dfrac{1}{2}$	$\beta \geq \dfrac{1}{4}$	Improved accuracy Provides artificial damping in higher modes Good for small Dt
$\alpha = 1/6$	$\beta = 1/2$	Improved accuracy Corresponds to a linear acceleration Very good for small Dt but tends to be unstable for large Dt

REMARKS

- In the direct integration approach, the equations of motion are integrated directly without special transformation. Such schemes may be categorized as explicit and implicit schemes.
- Explicit schemes are usually fast, efficient, and require no factorization of the effective stiffness matrix provided that the mass and damping are diagonal (refer to Equation 4.6 for an example). However, explicit schemes require the time step to be smaller than a critical value (Equation 4.7) that normally leads to very large number of steps.
- Implicit schemes require factorization of the effective stiffness matrix of the structure at each time step (refer to Equation 4.10 for an example) but do not require the condition of a critical time step.
- Although implicit schemes are unconditionally stable, accuracy is not always guaranteed without careful consideration of the step size (refer to Table 4.3).

4.4.3   Modal Superposition Method

In this method, the nodal displacement response is expressed in terms of the normal modes that may be found in an eigenvalue analysis. The coupled equations of motion, Equation 4.2, are first transformed into a set of independent or decoupled differential equations cast in modal generalized coordinates (Mirovitch, 1980). The dynamic response of the original system is then obtained by superimposing the responses of the uncoupled modal equations. The generalized coordinates (modal coordinates) are introduced by the following coordinate transformation

$$\mathbf{u} = \mathbf{\Phi}\mathbf{q} \text{ and } u_i = \sum_{j=1}^{m} \varphi_{ij}q_j \tag{4.11}$$

where $\mathbf{\Phi}$ is an $n \times m$ matrix called the eigenvector or mode shape matrix, m is the number of eigenvectors ($m \leq n$), n is the number of DoF of the system, and $\mathbf{q}$ is a vector of size $m \times 1$ representing the number of mode-amplitude generalized or normal coordinates. The transformation given by Equation 4.11 represents a change of basis from nodal displacements ($\mathbf{u}$) to modal generalized coordinates ($\mathbf{q}$). In order for the equations to be decoupled in the modal generalized coordinate system, the triple product $\mathbf{\Phi}^{\mathrm{T}}\mathbf{C}\mathbf{\Phi}$ has to be a diagonal matrix and the following orthogonality property is assumed for damping:

$$\mathbf{\Phi}^{\mathrm{T}}\mathbf{C}\mathbf{\Phi} = \mathrm{diag}(2\xi_i\omega_i) \tag{4.12}$$

where ξ_i is the damping ratio for mode i and $\mathrm{diag}(2\xi_i\omega_i)$ indicates a diagonal matrix with the ith diagonal component of $2\xi_i\omega_i$. This form has been assumed by generalization of damping for a single DoF system. It is practically more convenient to define the damping by the damping ratios of each mode than it is to evaluate the damping matrix explicitly. Introducing Equation 4.11 and Equation 4.12 into the equation of motion, the following decoupled equations of motion are obtained:

$$\ddot{\mathbf{q}} + \mathrm{diag}(2\xi_i\omega_i)\dot{\mathbf{q}} + \mathrm{diag}(\omega_i^2)\mathbf{q} = \mathbf{\Phi}^{\mathrm{T}}\mathbf{p}(t) \tag{4.13}$$

If the components of the applied force vector have the same time function, so that $\mathbf{p}(t)$ can be expressed as

$$\mathbf{p}(t) = \bar{\mathbf{p}}g(t) \tag{4.14}$$

where $\bar{\mathbf{p}}$ is a constant vector and $g(t)$ is a function giving the time change of the load vector, then the modal load can be written as

$$f_r(t) = (\varphi_r^{\mathrm{T}}\bar{\mathbf{p}})g(t) = \Gamma_r g(t) \tag{4.15}$$

where φ_r^{T} is a row modal vector corresponding to mode r. The quantity Γ_r is referred to as the modal participation factor (*NISA User's Manual*, 1992). A particular use of this quantity is in ground motion types of excitation.

Once the generalized coordinates, $\mathbf{q}$, are evaluated, the physical response of the original system in terms of nodal displacements, velocities, accelerations, and stresses are recovered from Equation 4.11. For the stress recovery, it is noted that (*NISA User's Manual*, 1992; Kim et al., 1994):

$$\boldsymbol{\sigma}(t) = \mathbf{D}\mathbf{B}\bar{\mathbf{u}}^{(e)}(t) = \mathbf{D}\mathbf{B}\,\boldsymbol{\Phi}^{(e)}\mathbf{q} = [\mathbf{D}\mathbf{B}\,\boldsymbol{\Phi}^{(e)}]\,\mathbf{q} = \Lambda\mathbf{q} \quad \text{or} \quad \sigma_i(t) = \sum_{r=1}^{m} \Lambda_{ir}q_r \tag{4.16}$$

where $\sigma_i(t)$ represents a stress component in an element (i.e., σ_{xx}, σ_{yy}, σ_{xy}, etc.), Λ_{ir} lists the stress components corresponding to the rth mode at a typical point. That is, Λ_{ir}, $r = 1, 2, \ldots, m$ are modal stresses which should be interpreted as stress shapes rather than absolute values of stress and the matrix $\mathbf{D}$ is the material stress–strain matrix.

The above derivation of the mode superposition method shows the decoupling advantage of the normal coordinates, whereby the change of basis from nodal displacements to normal modes yields a set of uncoupled modal equations, with each equation cast in the form of a single DoF oscillator. As indicated above, another major advantage of the normal mode method is that a good approximation to the response may be obtained using a drastically reduced number of coordinates ($m \ll n$). For most types of loading, with the exclusion of shock and impact loading, the contributions of the lowest few modes are generally more pronounced than the higher modes. Furthermore, practical finite element idealization approximates the lower modes better and tends to be less reliable for higher modes of vibration.

REMARKS

- In the modal superposition approach, the equations of motion are first transformed into a set of decoupled equations cast in modal generalized coordinates. The response of the system is then obtained by superimposing the solutions of the decoupled modal equations (Equation 4.13 and Equation 4.16).
- The modal superposition approach may only be achieved for linear systems with proportional or directly assumed modal damping (refer to Section 4.4.4).
- In the modal superposition approach, a good approximation to the response may be obtained using a drastically reduced number of modes. For most types of loading, with the exclusion of shock and impact loading, the contributions of the lowest few modes are generally more pronounced than the higher modes.
- Practical FE idealization approximates the lower modes better and tends to be less reliable for higher modes of vibration.

4.4.4 Damping Formulation

Damping is a source of energy dissipation in the structure and it leads to decay of the free vibration amplitude with time. Damping sources in a structure include internal friction in the material, Coulomb friction in sliding and pin joints, and other viscous friction forces due to fluid friction (Beards, 1982; Fretzen, 1986). The overall damping of a system is normally quite difficult to obtain and, in general, must be determined experimentally (Kareem and Gurley, 1996). In free vibrations and modal analysis, the damping may be neglected or an overall small value for the system may be assumed. This is a reasonable assumption since in practice damping is small enough and can usually be assumed to be viscous. In forced vibrations, however, and when the forcing frequency is close to one of the system's natural frequencies, the response of the system is dominated by the specified damping values.

Commonly, damping is described in viscous or structural form. Both descriptions are used for their mathematical convenience but may not truly represent the actual damping behavior and mechanism. For viscous damping, the damping force is proportional to and opposes the velocity. With structural damping, the damping force is proportional in magnitude to the internal elastic force (i.e., to the displacement) and is in the opposite direction to the velocity. In practical FE analysis, most programs provide the following damping representation that may be used for dynamic analysis:

1. Discrete viscous damper elements (dashpots). As discussed in Section 4.1.2, these elements are damping counterparts of the spring elements and are discrete idealization of viscous damping in the structure. The damping matrix resulting from these elements, in general, cannot be decoupled as in Equation 4.12. Hence, these elements may be used only with direct integration solution methods.

2. Proportional viscous damping (Rayleigh damping). In this type, the following arbitrary form for the damping matrix is assumed:

$$\mathbf{C} = c_1 \mathbf{K} + c_2 \mathbf{M} \tag{4.17}$$

where c_1 and c_2 are constants to be determined and supplied by the user. A commonly used method in determining the constants c_1 and c_2 is to define two damping rations ξ_1 and ξ_2 corresponding to two unequal natural frequencies ω_1 and ω_2, respectively. Since $\mathbf{C}$ is proportional to $\mathbf{K}$ and/or $\mathbf{M}$, it satisfies the orthogonality property and we have

$$c_1 \omega_i^2 + c_2 = 2\xi_i \omega_i \tag{4.18}$$

which leads to

$$\xi_i = \frac{c_1 \omega_i}{2} + \frac{c_2}{2\omega_i} \tag{4.19}$$

Equation 4.19 is then used to calculate the coefficient c_1 and c_2 from the knowledge of the specified damping rations ξ_1 and ξ_2 and their corresponding natural frequencies ω_1 and ω_2. For the case when $\mathbf{C}$ is only proportional to $\mathbf{K}$ ($c_2 = 0$), the damping ratio is proportional to the natural frequency, and thus the higher modes will be damped more than the lower modes. Similarly, if $\mathbf{C}$ is only proportional to $\mathbf{M}$ ($c_1 = 0$), the damping ratio is inversely proportional to the natural frequency and the higher modes will have less damping than the lower modes. The relationship of Equation 4.19 is diagrammatically illustrated in Figure 4.24.

The above method for calculating the coefficients c_1 and c_2 should not restrict the use of proportional damping to only modal analysis. Proportional damping may be used in harmonic, modal, and transient analyses, as well as substructure and reduced types of analyses.

For direct transient dynamic analysis, different sets of constants c_1 and c_2 may be assigned to different parts of the model. This may result in a nonorthogonal global damping matrix. The damping orthogonality condition, however, is not required in direct transient dynamic analysis since the governing equations are directly integrated.

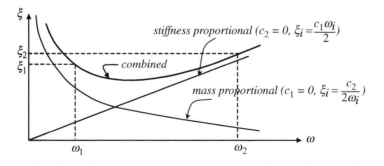

FIGURE 4.24 Proportional damping.

TABLE 4.4 Typical Damping Ratios for Various Systems and Materials

System/Material	Damping Ratio
Various metals in elastic range	<0.01
Small diameter piping systems	0.01–0.02
Large building during earthquake	0.01–0.05
Large diameter piping systems	0.02–0.03
Welded joints and rigid metal structures	0.02–0.04
Prestressed concrete structures	0.02–0.05
Metal structures with joints	0.03–0.07
Transmission lines (aluminum or steel)	0.04
Reinforced concrete structures	0.04–0.07
Rubber	0.05
Bolted joints	0.07
Shock absorbers	0.30

3. Modal viscous damping, in which the damping ratios are directly specified for the participating modes. This type of damping can be used in modal dynamic analyses and usually provides good results for systems with small damping. In this type, the damping matrix is assumed to be diagonal with values of $2\xi_i\omega_i$, where i is the ith diagonal and ξ_i is the damping ratio for mode i with frequency ω_i.

Typical values of overall damping ratios for various systems are given in Table 4.4 (Adams and Askenazi, 1999).

REMARKS

- Damping is a source of energy dissipation in the structure and is generally difficult to quantify. Damping sources in a structure include internal, Coulomb, and viscous friction.
- The most common forms of damping provided by FE programs are discrete, proportional, and modal viscous damping.
- Discrete damping is modeled by special spring and damping elements in FE applications. The damping matrix resulting from these elements cannot, in general, be decoupled as in Equation 4.12. Hence, these elements may be used only with direct integration solution methods.
- Proportional viscous or Rayleigh damping (Equation 4.17) leads to a decoupled damping matrix and may be used in both direct integration as well as modal superposition approaches.
- A commonly used method to determine the coefficient of proportionality is through defining two modal damping ratios (Equation 4.19).
- In modal viscous damping, the damping ratios are directly specified for the participating modes and the damping matrix is assumed to be diagonal.

4.5 Various Dynamic Analyses

4.5.1 Modal Analysis

This is normally the first, and in certain cases the only, type of analysis required for a structure. It is used to obtain the dynamic characteristics of a system in terms of its natural frequencies or eigenvalues and associated undamped free vibration natural modes or eigenvectors. These eigenmodes may then be used to decouple the general equations of motion in other types of dynamic analyses. Modal analysis is the first step required before performing mode superposition harmonic or transient analysis as well as spectrum analysis. The natural frequencies of the system may be obtained by solving the eigenvalue problem:

$$(\mathbf{K} - \omega^2 \mathbf{M})\bar{\varphi} = 0 \tag{4.20}$$

For a nontrivial solution, the coefficient matrix $(\mathbf{K} - \omega^2 \mathbf{M})$ is singular and its determinant is equal to zero, which yields the values of the natural frequencies. For each eigenvalue, ω_{n_i}, a corresponding mode shape, $\bar{\varphi}_i$, can be obtained from Equation 4.20. If the mass matrix has some zero diagonal values, the number of eigenpairs will be less than the total number of DoF by the number of zero entries on the diagonal of the mass matrix. For properly constrained structures, $\mathbf{K}$ is positive definite and all natural frequencies are positive. If the structure is unconstrained or partially constrained, however, $\mathbf{K}$ is positive semidefinite and the eigenvalues will contain zero frequencies representing the rigid-body modes. The calculated mode shapes from Equation 4.20 are normalized and they satisfy the orthogonality properties: $\varphi_i^{\mathrm{T}}\mathbf{M}\varphi_j = 0$ and $\varphi_i^{\mathrm{T}}\mathbf{K}\varphi_j = 0$ for $i \neq j$ and $\varphi_i^{\mathrm{T}}\mathbf{M}\varphi_i = 1.0$ and $\varphi_i^{\mathrm{T}}\mathbf{K}\varphi_i = \omega_i^2$, where $\varphi_i = \bar{\varphi}_i/\sqrt{m_i}$ and $m_i = \bar{\varphi}_i^{\mathrm{T}}\mathbf{M}\bar{\varphi}_i$.

Several eigenvalue extraction methods are available. Among the most commonly used are the block Lanczos, Subspace Iteration, Power Dynamics, and Reduced Householder methods (Mirovitch, 1980; Belytschko and Hughes, 1983; Bathe, 1996). The Lanczos method is recommended for large models with many modes to be extracted (about 50 or more). It is also better suited for models containing different types of elements such as solid, shell, and beam elements with the possibility that some elements are distorted. The method is fast and efficient but requires more memory than the Subspace Iteration method. In contrast, the Subspace Iteration method is well suited for extracting lower numbers of modes (less than 50) in models with well-shaped elements and it requires less memory than the Lanczos method. The Power Dynamics method is used for very large models, with more than 100,000 DoF, and is especially useful for obtaining a solution for the first several modes of the model. The Reduced Householder method is recommended for finding all the modes in small to medium models of less than 10,000 DoF (*ANSYS Users Manual*, 2003).

The information obtained from an eigenvalue or modal analysis is vital in the analysis of dynamic systems. The designer's goal is to ensure that the loading frequency is not close to any natural frequency of the system and that the mode shapes do not produce excessive deformation in weak sections of the structure. With respect to the mode shapes, it is desirable to avoid mode shapes that are similar to deformation patterns obtained from the static loading on the structure. As stated previously, the comfortable range for an operating frequency is three times higher than the nearest natural frequency from the lower side and three times smaller than the nearest natural frequency from the higher side. The natural frequencies may be altered by changing the mass distribution or adjusting the geometry of the system. If the loading frequency is higher than one or more of the first natural frequencies then large vibrations or "shudder" may occur during startup and in certain cases special startup procedures may have to be devised. If the loading frequency is close to one of the natural frequencies, a large factor of safety should be used. For example, an operating frequency 20% away from a natural frequency may call for doubling the nominal loads applied on a shaft. In some very special applications, it may be desirable actually to operate the system at one of its natural frequencies and utilize the amplification in the response, e.g., ultrasonic welding equipment and certain nanoscale measuring instruments.

Modal analysis requires no special boundary conditions and the structure may be completely free or partially constrained. Special attention should be given to the application of constraints, since overconstraining a system will result in overprediction of the first modes, which are generally nonconservative in a modal analysis.

REMARKS

- Modal dynamic analysis is used to obtain the dynamic characteristics of a system in terms of its natural frequencies (eigenvalues) and natural modes (eigenvectors). It is normally the first (and may be the only) dynamic analysis to be performed and its results are required to decouple the equations of motion for other analyses.
- Common eigenvalue extraction methods include: Lanczos (recommended for large models with many modes to be extracted; about 50 or more); Subspace Iteration (recommended for extracting lower number of modes, less than 50); Power Dynamics (recommended for very large models of more than 100,000 DoF); and Householder (recommended to find all modes in small to medium sized models of less than 10,000 DoF).
- Modal analysis requires no special boundary conditions and the structure may be completely free.

4.5.2   Transient Dynamic Analysis

4.5.2.1   Direct Integration Method

As indicated above, with the direct integration method, the general equation of motion is integrated directly without transformation. The input loads are specified as a function of time and the response is calculated in the time domain. Transient analysis could be viewed as solving an equivalent static analysis problem for each time step in the domain. The response quantities may include the time history of displacements, velocities, and accelerations at each point in the structure. Most programs will also provide a stress history output that is calculated in a postprocessing phase. The output results are straightforward and easy to interpret. As in static analysis, structures undergoing transient analysis should be fully constrained and all rigid-body motion should be eliminated. Initial conditions of displacement and velocity are also required to perform the analysis.

As discussed in Section 4.4.2, direct integration may be performed through explicit or implicit schemes. It should be noted that while implicit schemes do not require a critical time step, the size of the time step will affect the accuracy of the solution. Most programs are capable of automatic time stepping, which should be seriously considered. If automatic time stepping is not available in the program, the rule of thumb is to use a time step that is at least one tenth of the load period. Smaller time steps may be required to capture peak response and sharp changes in the input load.

The equations of motion (Equation 4.2) may first be reduced in size before the start of the integration process by choosing certain master DoF as discussed above. The reduced equations are then solved by one of the direct integration schemes and the results may then be expanded to the full set of DoF.

The following steps may be used as general guidelines in performing direct integration transient dynamic analysis:

- *Build the FE model*: This will include identifying the problem type and the type(s) of elements to be used in the model. The mesh should be fine enough to capture the highest mode required for the analysis. If stresses are required, the intensity of the mesh should be adequate for stress calculations.
- *Apply the loads and boundary conditions*: All rigid-body modes should be constrained and the model should be fully constrained. Time-dependent loads are normally specified in a table format.
- *Specify initial conditions*: Initial conditions of displacement and velocity are required for transient analysis. Most programs assume zero initial conditions for acceleration. If initial accelerations are not zero, the analyst may apply appropriate acceleration loads over a small time interval. It should be noted that initial conditions are required for all unconstrained DoF.

- *Set solution control parameters, start solution and review results*: This includes mass and damping formulation, damping parameters, time integration, and time stepping parameters as well as parameters to control output and postprocessing options.

4.5.2.2 Mode Superposition Method

In this method, the uncoupled modal equation 4.13, repeated here for convenience,

$$\ddot{\mathbf{q}} + \text{diag}(2\xi_r\omega_r)\dot{\mathbf{q}} + \text{diag}(\omega_r^2)\mathbf{q} = \{f_r(t)\} = \mathbf{\Phi}^T\mathbf{p}(t) \tag{4.21}$$

consist of m independent second-order differential equations with constant coefficients, where m is the number of the participating modes used in the mode superposition process.

At time $t = 0$, each modal equation is subject to the initial conditions $q_r(0) = \varphi_r^T\mathbf{Mu}_0$ and $\dot{q}_r(0) = \varphi_r^T\mathbf{M\dot{u}}_0$. For the underdamped case ($\xi_r < 1$), the solution to a typical modal equation (excluding rigid-body modes) is

$$q_r(t) = e^{-\xi_r\omega_r t}[\alpha_r \sin \bar{\omega}_r t + \beta_r \cos \bar{\omega}_r t] + \int_0^t f_r(\tau)h_r(t - \tau)d\tau \tag{4.22}$$

where $h_r(t - \tau)$ is the unit-impulse response function, defined by

$$h_r(t - \tau) = \frac{1}{\bar{\omega}_r}e^{-\xi_r\omega_r(t-\tau)} \sin \bar{\omega}_r(t - \tau) \tag{4.23}$$

The first term in Equation 4.23 represents the free vibration response (homogenous solution) and the second term represents the particular solution (Duhamel integral). α_r and β_r are constants evaluated from the initial conditions and $\bar{\omega}_r$ is the damped natural frequency given by

$$\bar{\omega}_r = \omega_r\sqrt{1 - \xi_r^2} \tag{4.24}$$

The integral in Equation 4.22 may be evaluated in closed form if the applied loading $\mathbf{p}(t)$, and consequently $f_r(t)$, is a step input, ramp input, piecewise linear function of time, or harmonic input; otherwise, numerical integration is utilized to obtain the response.

The general steps of a typical modal transient analysis include building the model, extracting the required modes for the analysis, obtaining the modal transient analysis and, finally, expanding the modal superposition solution. It should be noted that most programs do not allow change of displacement constraints after the mode extraction step is performed; i.e., all displacement constraints should be specified before performing the modal analysis to extract the required modes. Some programs may require the full loading conditions to be specified in the modal extraction step. Initial conditions should be specified before performing the modal transient solution. This may include initial displacement and velocities.

REMARKS

- In transient analysis, the input loads are specified as a function of time and the response (displacement, velocity, acceleration, and stress) is calculated in the time domain. The structure should be fully constrained and initial conditions are required.
- Direct integration transient analysis may be performed with explicit (conditionally stable) or implicit (unconditionally stable) schemes.
- Prior to performing integration, the size of the equations may be reduced by choosing master DoF and condensing out slave DoF.
- Transient analysis with mode superposition requires an initial modal analysis run followed by the superposition of modal responses (Equation 4.22) obtained from the solution of the decoupled equations. The integral in Equation 4.22 may be evaluated in closed form for step, ramp, piecewise linear, and harmonic inputs; otherwise, numerical integration is utilized.

4.5.3 Harmonic Response Analysis

4.5.3.1 Direct Integration Method

Harmonic response analysis involves computing the steady-state response of a structure to a set of harmonic concentrated loads, pressure loads, and harmonic ground motion. The harmonic loads may be defined in terms of different amplitude and phase spectra. In general, the load vector may be represented in the form:

$$p_i(t) = p_i(\Omega)\sin[\Omega t + \Psi_i(\Omega)] \tag{4.25}$$

where $p_i(t)$ is a component of the forcing function having a magnitude, $p_i(\Omega)$, phase shift, $\psi_i(\Omega)$, and forcing frequency, Ω.

Substituting Equation 4.25 into Equation 4.2, we obtain:

$$\mathbf{M\ddot{u} + C\dot{u} + Ku} = p(t) = \bar{p}(\Omega)\sin[\Omega t + \bar{\Psi}(\Omega)] \tag{4.26}$$

Using direct integration schemes to solve Equation 4.26 is theoretically viable but not recommended practically for a few reasons. A large number of time steps would be normally required to obtain a meaningful response to a harmonic excitation and the equations have to be resolved for each forcing frequency. Also, if the damping is zero and a particular forcing frequency coincides with one of the natural frequencies, the matrix becomes singular and the solution process fails. If the forcing frequency is close to one of the natural frequencies, the matrix becomes ill conditioned and poses solution problems even if the damping is not zero. For these reasons, it is recommended to perform harmonic response analyses using the mode superposition method.

4.5.3.2 Mode Superposition Method

In the mode superposition method, Equation 4.26 is transformed into generalized modal coordinates and decoupled. For the *r*th mode, the equation takes the form:

$$\ddot{\mathbf{q}}_r + 2\xi_r\omega_r\dot{\mathbf{q}}_r + \omega_r^2 q_r = \bar{f}_r e^{i\Omega t} \tag{4.27}$$

where $\bar{f}_r$ is the amplitude of modal load given by $\bar{f}_r = \varphi_r^{\mathrm{T}}\bar{\mathbf{p}}$. The steady-state modal response is given by

$$q_r(t) = \bar{q}_r e^{i\Omega t} \tag{4.28}$$

which when substituted into Equation 4.27 gives

$$\bar{q}_r(\Omega) = H_r(\Omega)\cdot\bar{f}_r(\Omega) \tag{4.29}$$

and

$$H_r(\Omega) = \cfrac{1}{\omega_r^2\left[\left(1 - \cfrac{\Omega^2}{\omega_r^2}\right) + i\left(2\xi_r\cfrac{\Omega}{\omega_r}\right)\right]} \tag{4.30}$$

Equation 4.29 in the frequency domain is equivalent to Equation 4.22 in the time domain and $H_r(\Omega)$ is the Fourier transform of the unit-impulse response function $h_r(t)$.

The physical response is recovered from the generalized modal response through the transformation $\bar{\mathbf{u}} = \mathbf{\Phi}\bar{\mathbf{q}}$. The velocity and acceleration components are obtained from the corresponding displacement component through multiplication by Ω and Ω^2, respectively. Finally, the stress components are obtained by the use of modal stresses (Equation 4.16). Most programs provide the option that the user can obtain the responses in either an amplitude–phase format or real–imaginary format and the actual value of all responses can also be obtained for a given value of Ωt.

The input to harmonic analysis constitutes the forcing frequency and the magnitude of the load vector at various points of load applications. For each loading case, all forces should have the same frequency. A change in the forcing frequency gives rise to a new load case. In harmonic analysis, as is the case in transient and static analyses, the structure should be fully constrained and all rigid-body motions should be eliminated. If the structure has rigid-body modes in one of the DoF, the analyst may use soft springs to ground the structure or utilize symmetry to provide constraints.

The general steps of a typical modal harmonic analysis are quite similar to those for the modal transient analysis, i.e., they include building the model, extracting the required modes for the analysis, obtaining the modal harmonic analysis and, finally, expanding the modal superposition solution. It should also be noted that most programs do not allow changes in displacement constraints after the mode extraction step, i.e., all displacement constraints should be specified before performing the modal analysis to extract the required modes. Some programs may require the full loading conditions to be specified in the modal extraction step.

Remarks

- Harmonic response analysis involves computing the steady-state response of a structure to a set of harmonic loads and ground motion (Equation 4.26).
- Using direct integration schemes in harmonic analysis is theoretically viable but not recommended or used practically.
- Harmonic analysis with mode superposition requires an initial modal analysis run followed by the superposition of modal responses (Equation 4.28) obtained from the solution of the decoupled equations.
- The input to harmonic analysis constitutes the frequency and the magnitude of the load vector at all points of load applications. For each load case, all forces should have the same frequency. The structure should be fully constrained.

4.5.4 Response Spectrum Analysis

This analysis is an efficient alternative to transient dynamic analysis for estimating the maximum response under support excitations without regard to the time at which the maximum occurs. If the input loading is classified as a shock or impulse, this analysis is termed "shock spectrum analysis." In theory, one may use a direct integration scheme to find the response as a function of time and then use the maximum value of the response. This may be prohibitive because of the very large number of time steps that will be required to capture the peak response. As a more efficient alternative, a modal analysis should be performed first and then a sufficient number of modes should be retained for the response spectrum analysis. Then the solution of the individual decoupled modal equation (Equation 4.13) for a viscous underdamped system under ground motion $w(t)$ is

$$q_r(t) = \frac{1}{\bar{\omega}_r} \int_0^t \ddot{w}(\tau) e^{-\xi_r \omega_r (t - \tau)} \sin[\bar{\omega}_r (t - \tau)] d\tau \qquad (4.31)$$

The response function in Equation 4.31 is scanned over time to find the maximum scalar value $(q_r)_{\max}$. Then the physical value of the response of the ith DoF due to $(q_r)_{\max}$ may be given by

$$\bar{u}_{ir} = \Phi_{ir}(q_r)_{\max} \qquad (4.32)$$

in which Φ_{ir} is the component of the modal vector, r, in direction i. Equation 4.32 is different from Equation 4.11 in which all the components of the modal vector, Φ_i, and the generalized response, q_i, are used and, therefore, actual and accurate physical displacements are obtained.

It can be readily seen that the quantities $(q_r)_{\max}$, and hence $\bar{u}_{ir}$, are only functions of the natural frequency and damping (in addition to the ground excitation). A shock (or response) spectrum curve for a certain value of damping may be defined as the maximum responses of all such single DoF systems with a given damping and plotted as a function of natural frequency. Similarly, the maximum acceleration and maximum velocity may also be determined and are termed as spectral acceleration and spectral velocity, respectively.

The mass matrix in response spectrum analysis may be lumped or modified lumped based on the consistent mass matrix. Response spectrum analysis using consistent mass generally lowers the resulting internal forces and moments in the structure and therefore produces less stresses and may be considered generally nonconservative (Gregory, 1990).

Unlike transient dynamic analysis, the contributions of the physical responses for each of the modes cannot be directly summed to obtain the total response. This is because the maxima for each mode occur at different times and the information on the time of maxima is not available in the shock spectra. Reasonable, but arbitrary, estimates of the maxima may be obtained by using one of the following commonly used modal combination methods (Wilson et al., 1981; *NISA User's Manual*, 1992; Roussel, 1994; Joshi and Gupta, 1998):

1. *The sum of absolute magnitudes (ABS or PEAK)*: The absolute sum of the modal responses is given by

$$R_{\text{tot}} = \sum_{r=1}^{m} |R_r| \qquad (4.33)$$

in which R_r is the physical response (displacement, velocity, or acceleration) due to mode r and m is the number of modes considered. This method is conservative and is used if the natural frequencies are closely spaced (within 10% of each other) and/or when damping is large. It is also shown that the method may lead to unrealistically high calculated responses, e.g., in the coupled analysis of light secondary systems attached to heavy primary structures or in the decoupled analysis of systems when the centers of mass and stiffness do not coincide (Mertens, 1994).

2. *The square root of the sum of the squares (SRSS or RMS)*: The square root of the sum of the squares of the modal response is given by

$$R_{\text{tot}} = \sqrt{\sum_{r=1}^{m} R_r^2} \qquad (4.34)$$

This method is applicable if the modes are statistically independent, which is the case when the natural frequencies are far apart and/or when damping is small. The SRSS method is commonly used in a wide range of applications. Some studies indicate, however, that the method may underestimate the response of structures with high-frequency modes, e.g., long-span bridges exposed to high wind velocities (Joshi and Gupta, 1998).

3. *The complete quadratic combination (CQC)*: The complete quadratic combination is given by the following formula (Wilson et al., 1981; Der Kiureghian and Nakamura, 1993):

$$R_{\text{tot}} = \sqrt{\sum_{r=1}^{m} \sum_{s=1}^{m} R_r \cdot R_s P_{rs}} \qquad (4.35)$$

in which

$$P_{rs} = \frac{8\sqrt{\xi_r \xi_s}(\xi_r + \gamma \xi_s)\gamma^{3/2}}{(1 - \gamma^2) + 4\xi_r \xi_s (1 + \gamma^2) + 4\gamma^2(\xi_r^2 + \xi_s^2)} \qquad (4.36)$$

where ξ_r and ξ_s are modal damping ratios and $\gamma = \omega_s/\omega_r$. This method encompasses the SRSS and ABS procedures for $\xi_r = \xi_s$. If $\gamma = 0$, the CQC method reduces to the former, and if $\gamma = 1$, it reduces to the latter. Certain modifications of the CQC method exist, aimed at improving response estimates for structures with high frequency modes (Der Kiureghian and Nakamura, 1993).

TABLE 4.5   Common Methods of Combining Modal Responses

Combinational Method	Comments
Sum of absolute magnitudes (ABS or PEAK)	Generally gives conservative results
	Used if the natural frequencies are closely spaced (within 10% of each other) and/or when damping is large
Square root of the sum of the squares (SRSS or RMS)	Use when natural frequencies are far apart and/or when damping is small
	May underestimate the response of structures with high frequency modes
Complete quadratic combination (CQC)	Combination of the SRSS and the ABS methods
	Certain modifications exist to correct the response of structures with high frequency modes
Combination of ABS and SRSS	Another form of combination of the SRSS and the ABS methods employed by simply adding the response of the two methods

4. *Combination of ABS and SRSS*: The absolute maxima of the modal responses is added to the square root of the sum of the squares of the remaining modal responses as follows (*NISA User's Manual*, 1992):

$$R_{\max} = |R_j| + \sqrt{\sum_{r=1, r \neq j}^{m} R_r^2} \tag{4.37}$$

where

$$R_j = \max \text{ for all } {}_r |R_r| \tag{4.38}$$

The above superposition rules are employed to get the response maxima due to ground excitation in one direction. Similarly, the maximum responses due to ground motions in other directions are computed separately. These maxima are then superimposed by using the SRSS or ABS procedures to get the total response.

A final note is given here on the use of spectrum analysis in earthquake-resistant design. Most building codes require the designer to consider the response of the structure due to simultaneous action of three translational components of ground motion: two in the horizontal plane and one in the vertical direction. Standard design practice is to calculate the peak responses of these inputs independently and then combine these peak responses according to rules similar to those discussed above (Menun and Der Kiureghian, 1998; Lopez et al., 2001).

Table 4.5 summarizes the above modal combination methods, their use and merits.

REMARKS

- Response spectrum and shock analysis are efficient alternatives to transient dynamic analysis for estimating the maximum response under support excitations without regard to the time at which the maximum occurs.
- In response spectrum analysis, the solution of the individual decoupled modal equation (Equation 4.13) for ground motion, $w(t)$, is obtained (Equation 4.31) and the maximum value of the response is recorded (with no reference to the time it occurs).
- Reasonable, but arbitrary, estimates of the maxima for the overall structure may be obtained by using various methods of combining the modal maximums (Table 4.5).

4.6   Checklist for Dynamic FE Analysis

1. Study the physical system and decide on the type of problem, i.e., beam, 2-D, shell, or 3-D, and the type of elements to be used. Typical questions to be addressed are:
   - Can the geometry be idealized as 2-D or shell or do we have to use 3-D analysis? (Section 4.1.2 to Section 4.1.6).
   - Can the loading and constraints be idealized as 2-D?
   - Are we concerned about responses that are not compatible with the loading type?
   - Are we interested in displacements, stresses, or natural frequencies?
   - Can we have a simple spring/mass model?
   - What type of element might be used in the analysis, i.e., linear, quadratic, or cubic (Section 4.1.7)? Generally, linear and quadratic elements are used for the majority of analyses. The choice between linear and quadratic elements is rather arbitrary and may depend on certain advanced analysis aspects. For linear analysis, there is little difference between the response of the two elements if the overall model has the same or similar total DoF.

2. Study all types of loading to identify the required analysis type. Important questions to be addressed are:
   - Is the load steady, cyclic, transient, or random (Section 4.2)?
   - What is the frequency content of the load? A simple Fourier analysis may be required to define the main harmonics of a transient load input.
   - What are the first few natural frequencies of the structure? This question will be answered in detail by performing a modal analysis. It is important, however, to have an approximate idea of the range of frequencies of the structure and to compare this to the loading frequency in order to identify the proper type of analysis (Section 4.2). In order to answer this question adequately, a very simplified mass/spring model and quick hand calculations may be considered.
   - What kind of data is required from the analysis? Data required may be in terms of displacements, velocities, accelerations, stresses, strains, etc. Also, identify if peak values or time histories are required. This decision may have an impact on what analysis type is required.
   - What is the type of analysis required (Section 4.2)?

3. Identify other data required to perform the analysis. This may include:
   - Material properties.
   - Damping consideration, modeling, and values (Section 4.4.4).
   - Identifying all load cases and all constraints of the system. Special attention must be paid to the modeling of constraints and nonstructural masses (Section 4.3).
   - If modal superposition is going to be used, how many modes are required for the analysis?

4. Start the actual modeling process, testing and verification of the model. Important issues to be considered may include:
   - A decision should be made of whether or not a simple or coarse model is required. In general, dynamic analysis of large models is quite costly. Dynamic analysis results (such as natural frequencies, mode shapes, and to some extent displacements, velocities, and accelerations) may be accurately obtained from a coarse model or even a mass/spring model. The main concern would be in obtaining stress and strain responses. In many cases, it may be advisable to have two different models — one for dynamic analysis and the other for stress analysis. The output from the dynamic model may be used as input to the stress model. In any case, if the model is large, it is advisable to have a coarser model for testing and basic understanding of the behavior of the structure (Section 4.3).
   - Choose a program to perform the analysis. The key issue in choosing a program would be the availability of the required analysis type. Other issues that may be considered include ease of model generation, ease of postprocessing, and cost effectiveness. It is always advisable to have the same program for all phases of the analysis, i.e., modeling, solution, and postprocessing.

- Build the model geometry (Section 4.1.1). In 2-D problems, the model geometry may be imported from a CAD program. In 3-D and shell problems, however, this process entails many problems and it may be faster for the analyst to build the model directly in a FE database.
- Build the finite element model and apply load cases and constraints (Section 4.2).
- Test the model. Most FE programs provide extensive testing capabilities that may be invoked to test the quality of the mesh and the continuity of the model. It is important to note that additional testing is required. The response of the system to simple load cases should be considered. Testing should involve load cases that may be easily verified. This may include computing structure response due to unit impulse, step input, or sudden release of an initial displacement. Also, simple static analysis runs can be very useful in this regard and may provide initial conditions for the final dynamic run.

5. Analyze the model for the actual load cases. The following important issues should be considered:
   - Identify solution parameters. This includes mass and damping formulation, damping parameters (Section 4.4.4), time integration, and time stepping parameters, as well as parameters to control output and postprocessing options. If modal analysis is considered, identify the modal extraction method (Section 4.5).
   - Check and apply restart capabilities in the program. Large size dynamic analyses should, in general, be done in steps and the restart option should be invoked.
   - If modal analysis is performed, a careful check of mode shapes will provide important information about the validity of the model and the appropriateness of solution steps to follow.
   - Identify solution control parameters that invoke the storage of specific data required for postprocessing.

6. Postprocess the results. This is the final stage of the analysis and may involve:
   - Graphical and contour plots of various response quantities. This may be done in the form of time history plots, envelopes of maximum responses, or contour plots of stress maxima. The amount of output data associated with large dynamic analysis runs is vast and such graphical display of results is essential.
   - Most programs have extensive capabilities in results animation, search facilities, preparing output summaries, and some can also assist in report writing. Such capabilities may become important in processing the results of large problems.

References

Adams, V. and Askenazi, A. 1999. *Building Better Products with Finite Element Analysis*, On Word Press, Santa Fe.

ANSYS User's Manual, Revision 7.0, Swanson Analysis Systems Inc., Houston, 2003.

Babuska, I. and Guo, B.Q., Approximation properties of the *h-p* version of the finite element method, *Comput. Methods Appl. Mech. Eng.*, 133, 319–346, 1996.

Bathe, K.J. 1996. *Finite Element Procedures in Engineering Analysis*, Prentice Hall, New York.

Bathe, K.J. and Dvorkin, E.N., A four-node plate bending element based on Mindlin–Reissner plate theory and a mixed interpolation, *Int. J. Numer. Methods Eng.*, 21, 367–383, 1985.

Beards, C.F., Damping in structural joints, *Sound Vib. Dig.*, 14, 9–11, 1982.

Belytschko, B. and Hughes, T.J.R., Eds. 1983. *Computational Methods for Transient Analysis*, North Holland, Amsterdam.

Boresi, A.P. and Chong, K.P. 2000. *Elasticity in Engineering Mechanics*, 2nd ed., Wiley, New York.

Bouhaddi, N. and Fillod, R., Method for selecting master DoF in dynamic substructuring using the Guyan condensation method, *Comput. Struct.*, 45, 941–946, 1992.

Cesar de Sa, J.M.A., Natal, J.R.M., Fontes, R.A., and Almeida, P.M., Development of shear locking-free shell elements using an enhanced assumed strain formulation, *Int. J. Numer. Methods Eng.*, 53, 1721–1750, 2002.

Chan, H.C., Cai, C.W., and Cheung, Y.K., Convergence studies of dynamic analysis by using the finite element method with lumped mass matrix, *J. Sound Vib.*, 165, 193–207, 1993.

Cook, R.D., Malkus, D.S., and Plesha, M.E. 1989. *Concepts and Applications of Finite Element Analysis*, 3rd ed., Wiley, New York.

Deiters, T.A. and Smith, K.S., Faster, easier finite element model reduction, *Exp. Tech.*, 24, 35–40, 2000.

Der Kiureghian, A. and Nakamura, Y., CQC modal combination rule for high-frequency modes, *Earthquake Eng. Struct. Dyn.*, 22, 943–956, 1993.

Freed, A.M. and Flanigan, C., Comparison of test-analysis model reduction methods, *Sound Vib.*, 25, 30–35, 1991.

Fretzen, C.P., Identification of mass, damping, and stiffness matrices of mechanical systems, *ASME J. Vib. Acoust. Stress Reliab. Des.*, 108, 9–16, 1986.

Gregory, F.M., Response spectra analysis. Consistent mass versus lumped mass, *ASME-Pressure Vessels Piping Div. (Publ.) PVP Seismic Eng.*, 197, 85–89, 1990.

Jensen, M.S., High convergence order finite elements with lumped mass matrix, *Int. J. Numer. Methods Eng.*, 39, 1879–1888, 1996.

Joshi, R.G. and Gupta, I.D., On the relative performance of spectrum superposition methods considering modal interaction effects, *Soil Dyn. Earthquake Eng.*, 17, 357–369, 1998.

Kamel, H.A., Integration of solid modeling and finite element generation, *Comput. Methods Appl. Mech. Eng.*, 89, 485–496, 1991.

Kareem, A. and Gurley, K., Damping in structures: its evaluation and treatment of uncertainty, *J. Wind Eng. Ind. Aerodyn.*, 59, 131–157, 1996.

Kim, K., Review of mass matrices for eigenproblems, *Comput. Struct.*, 46, 1041–1048, 1993.

Kim, H.M., Bartkowicz, T.J., and Van Horn, D.A., Data recovery and model reduction methods for large structures, *Finite Elem. Anal. Des.*, 16, 85–98, 1994.

Lopez, O.A., Chopra, A.K., and Hernandez, J.J., Evaluation of combination rules for maximum response calculation in multicomponent seismic analysis, *Earthquake Eng. Struct. Dyn.*, 30, 1379–1398, 2001.

Luo, Y., Explanation and elimination of shear locking and membrane locking with field consistence approach, *Comput. Methods Appl. Mech. Eng.*, 162, 249–269, 1998.

Menun, C. and Der Kiureghian, A., Replacement for the 30%, 40%, and SRSS rules for multicomponent seismic analysis, *Earthquake Spectra*, 14, 153–164, 1998.

Mertens, P.G., New algebraic white-noise modal combination rule — GAC(A), *Nucl. Eng. Des.*, 147, 299–309, 1994.

Mirovitch, L. 1980. *Computational Methods in Structural Dynamics*, Sijthoff and Noordhoff, Alphen a/d Rijn.

NISA User's Manual, Engineering Mechanics Research Corp., Michigan, USA, 1992.

Roussel, G., Algebraic modal combination method in seismic response spectrum analysis, *ASME-Pressure Vessels Piping Div. (Publ.) PVP Seismic Eng.*, 275-1, 211–216, 1994.

Wilson, E.L., Der Kiureghian, A., and Bayo, E.P., A replacement for the SRSS method in seismic analysis, *Earthquake Eng. Struct. Dyn.*, 9, 187–194, 1981.

Zeng, L.F., Wiberg, N.-E., and Bernspang, L., Adaptive finite element procedure for 2D dynamic transient analysis using direct integration, *Int. J. Numer. Methods Eng.*, 34, 997–1014, 1992.

5

Vibration Signal Analysis

Clarence W. de Silva
The University of British Columbia

Summary

This chapter considers the nature and analysis of vibration signals. Both time-domain techniques and frequency-domain techniques are investigated, which are related through Fourier transform. Topics covered include signal types, signal sampling, aliasing error, truncation error, window functions, spectral analysis, bandwidth issues, and order analysis. Several applications of signal analysis in mechanical vibration are discussed.

5.1   Introduction

Numerous examples can be drawn from engineering applications for vibrating (dynamic) systems. A steam generator of a nuclear power plant that undergoes flow-induced vibration; a high-rise building subjected to seismic motions at its foundation; an incinerator tower subjected to aerodynamic disturbances; an airplane excited by atmospheric turbulence; a gate valve under manual operation; and a heating, ventilating, and air conditioning (HVAC) control panel stressed due to vibrations in its support structure are such examples.

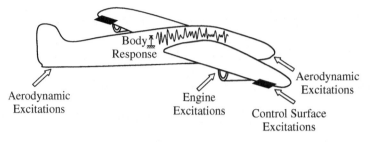

FIGURE 5.1   In-flight excitations and responses of an aircraft.

Consider an aircraft in flight, as schematically shown in Figure 5.1. There are many excitations on this dynamic system. For example, jet engine forces and control surface movements are intentional excitations, whereas aerodynamic disturbances are unintentional (and unwanted) excitations. The primary response of the aircraft to these excitations will be the motions in various degrees of freedom (DoF), including rigid-body and flexible (vibratory) mode motions.

Even though the inputs and outputs (excitations and responses) are functions of *time*, they can also be represented as functions of *frequency*, through *Fourier transformation*. The resulting *Fourier spectrum* of a signal can be interpreted as the set of frequency components which the original signal contains. This *frequency-domain* representation of a signal can highlight many salient characteristics of the signal and also those of the corresponding system. For this reason, frequency-domain methods, particularly Fourier analysis, are used in a wide variety of applications such as data acquisition and interpretation, experimental modeling and modal analysis, diagnostic techniques, signal/image processing and pattern recognition, acoustics and speech research, signal detection, telecommunications, and dynamic testing for design development, quality control, and qualification of products. Many such applications involve the study of mechanical vibrations.

5.2   Frequency Spectrum

Excitations (inputs) to a dynamic system progress with time, thereby producing responses (outputs) which themselves vary with time. These are signals that can be recorded or measured. A measured signal is a *time history*. Note that in this case the independent variable is *time* and the signal is represented in the *time domain*. A limited amount of information can be extracted by examining a time history. As an example, consider the time-history record shown in Figure 5.2. It can be characterized by parameters such as the following:

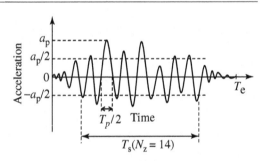

FIGURE 5.2   A time-history record.

a_p = peak amplitude
T_p = period in the neighborhood of the peak = 2 × interval between successive zero crossings near the peak
T_e = duration of the record
T_s = duration of strong response (i.e., the time interval beyond which no peaks occur that are larger than $a_p/2$)
N_z = number of zero crossings within T_s (N_z = 14 in Figure 5.2)

It is obviously cumbersome to keep track of so many parameters and, furthermore, not all of them are equally significant in a given application. Note, however, that all the parameters listed above are directly

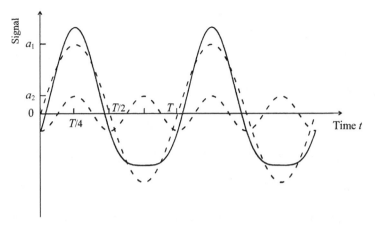

FIGURE 5.3 Time-domain representation of a periodic signal.

or indirectly related to either the *amplitude* or the frequency of zero crossings within a given time interval. This signifies the importance of a frequency variable in representing a time signal. This is probably the fundamental motivation for using frequency-domain representations. In this context, however, more rigorous definitions are needed for the parameters: amplitude and frequency. A third parameter, known as *phase angle*, is also needed for unique representation of a signal in the frequency domain.

5.2.1 Frequency

Let us further examine the basis of frequency-domain analysis. Consider the *periodic* signal of period T that is formed by combining two *harmonic* (or sinusoidal) components of periods T and $T/2$ and amplitudes a_1 and a_2 as shown in Figure 5.3. The *cyclic frequency* (cycles/sec, or hertz, or Hz) of the two components are $f_1 = 1/T$ and $f_2 = 2/T$. Note that in order to obtain the *angular frequency* (radians/sec), the cyclic frequency has to multiplied by 2π.

5.2.2 Amplitude Spectrum

An alternative graphical representation of the periodic signal shown in Figure 5.3 is given in Figure 5.4. In this representation, the amplitude of each harmonic component of the signal is plotted against the corresponding frequency. This is known as the *amplitude spectrum* of the signal, and it forms the basis of the frequency-domain representation. Note that this representation is often more compact, and can be far more useful than the time-domain representation. Note further that in the frequency-domain representation, the independent variable is frequency.

5.2.3 Phase Angle

In its present form, Figure 5.4 does not contain all the information of Figure 5.3. For instance, if the high-frequency component in Figure 5.3 is shifted through half its period ($T/4$), the resulting signal is shown in Figure 5.5. This signal is quite different from that in Figure 5.3 but since the amplitudes and the frequencies of the two harmonic components are identical for both signals, they possess the same amplitude spectrum. So, what is lacking in Figure 5.3 in order to make it a unique

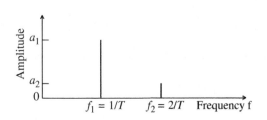

FIGURE 5.4 The amplitude spectrum of a periodic signal.

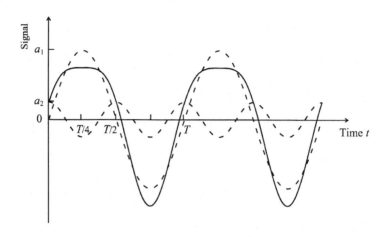

FIGURE 5.5   A periodic signal with an identical amplitude spectrum as for Figure 5.2.

representation of a signal, is the information concerning the exact location of the harmonic components with respect to the time reference or origin ($t = 0$). This is known as the phase information. For example, the distance of the first positive peak of each harmonic component from the time origin can be expressed as an angle (in radians) by multiplying it by $2\pi/T$. This is termed the *phase angle* of the particular component. In both signals (shown in Figure 5.3 and Figure 5.5) the phase angle of the first harmonic component is the same and equals $\pi/2$ according to the present convention. The phase angle of the second harmonic component is $\pi/2$ in Figure 5.3 and zero in Figure 5.5.

5.2.4   Phasor Representation of Harmonic Signals

A convenient geometric representation of a harmonic signal of the form

$$y(t) = a \cos(\omega t + \phi) \tag{5.1}$$

is possible by means of a phasor. This representation is illustrated in Figure 5.6. Specifically, consider a rotating arm of radius a, rotating in the counter-clockwise (ccw) direction at an angular speed of ω rad/sec. Suppose that the arm starts (i.e., at $t = 0$) at an angular position ϕ with respect to the y-axis (vertical axis) in the ccw sense. Then, it is clear from Figure 5.6(a) that the projection of the rotating arm on the y-axis gives the time signal $y(t)$. This is the phasor representation, where we have

Signal amplitude = length of the phasor
Signal frequency = angular speed of the phasor
Signal phase angle = initial position of the phasor with respect to the y-axis

It should be clear that a phase angle makes practical sense only when two or more signals are compared. This is so because for a given harmonic signal we can pick any point as the time reference ($t = 0$). However, when two harmonic signals are compared, as in Figure 5.6(b), we may consider one of those signals that starts (at $t = 0$) at its position peak as the reference signal. This will correspond to a phasor whose initial configuration coincides with the positive y-axis. As is clear from Figure 5.6(b), for this reference signal we have, $\phi = 0$. Then the phase angle ϕ of any other harmonic signal would correspond to the angular position of its phasor with respect to the reference phasor. Note that, in this example, the time shift between the two signals is ϕ/ω, which is also a direct representation of the phase. It should be clear, then, that the phase difference between two signals is also a representation of the *time lead* or *time lag* (delay) of one signal with respect to the other. Specifically, a phasor that is ahead of the reference phasor is considered to "lead" the reference signal. In other words, the signal $a \cos(\omega t + \phi)$ has a phase lead of ϕ or a time lead of ϕ/ω with respect to the signal of $a \cos \omega t$.

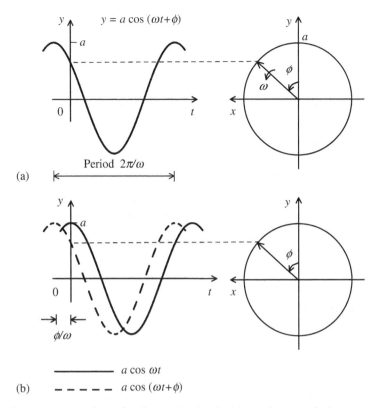

FIGURE 5.6 Phasor representation of a harmonic signal. (a) A phasor and the corresponding signal; (b) representation of a phase angle (phase lead) ϕ.

Another important observation may be made with regard to the phasor representation of a harmonic signal. A phasor may be expressed as the complex quantity

$$y(t) = a\,e^{j(\omega t + \phi)} = a\cos(\omega t + \phi) + ja\sin(\omega t + \phi) \tag{5.2}$$

whose real part is $a\cos(\omega t + \phi)$, which is in fact the signal of interest. It is clear from Figure 5.6 that, if we take the y-axis to be real and the x-axis to be imaginary, the complex representation 5.2 is indeed a complete representation of a phasor. By using the complex representation 5.2 for a harmonic signal, significant benefits of mathematical convenience could be derived in vibration analysis. It suffices to remember that practical vibrations are "real" signals, and regardless of the type of mathematical analysis that is used, only the real part of a complex signal of the form 5.2 will make physical sense.

5.2.5 RMS Amplitude Spectrum

If a harmonic signal $y(t)$ is averaged over one period T, the negative portion cancels out the positive portion, giving zero. Consider a harmonic signal of angular frequency ω (or cyclic frequency f), phase angle ϕ, and amplitude a, as given by

$$y(t) = a\cos(\omega t + \phi) = a\cos(2\pi ft + \phi) \tag{5.3}$$

Its average (mean) value is

$$y_{\text{mean}} = \frac{1}{T}\int_0^T y(t)\,\mathrm{d}t = 0 \tag{5.4}$$

which can be verified by direct integration, while noting that

$$T = 1/f = 2\pi/\omega \qquad (5.5)$$

For this reason, mean value is not a measure of the "strength" of a signal in general. Now let us define the *root mean square* (RMS) value of a signal. This is the square root of the mean value of the square of the signal. By direct integration, it can be shown that for a sinusoidal (or harmonic) signal; the RMS value is given by

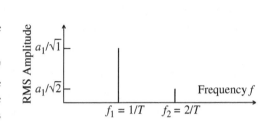
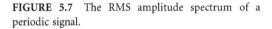

FIGURE 5.7 The RMS amplitude spectrum of a periodic signal.

$$y_{\mathrm{RMS}} = \left[\frac{1}{T} \int_0^T y^2(t)\mathrm{d}t \right]^{1/2} = \frac{a}{\sqrt{2}} \qquad (5.6)$$

It follows that the *RMS amplitude spectrum* is obtained by dividing the amplitude spectrum by $\sqrt{2}$. For example, for the periodic signal formed by combining two harmonic components as in Figure 5.3, the RMS amplitude spectrum is shown in Figure 5.7. This again is a frequency-domain representation of a signal, and the independent variable is frequency.

5.2.6   One-Sided and Two-Sided Spectra

Mean squared amplitude spectrum of a signal (sometimes called *power spectrum* because the square of a variable such as voltage and velocity is a measure of quantities such as *power* and *energy*, even though it is not strictly the spectrum of power in the conventional sense) is obtained by plotting the mean squared amplitude of the signal against frequency. Note that these are *one-sided spectra* because only the positive *frequency band* is considered. This is a realistic representation because one cannot talk about negative frequencies for a real system. But, from a mathematical point of view, we may consider negative frequencies as well. In a spectral representation it is at times convenient to consider the entire frequency band (consisting of both negative and positive values of frequency). It then becomes a *two-sided spectrum*. In this case the spectral component at each frequency value should be equally divided between the positive and the negative frequency values (hence, the spectrum is *symmetric*), such that the overall mean squared amplitude (or power or energy) remains the same.

We have seen that for a harmonic signal component of amplitude a and frequency f (e.g., $a \cos(2\pi f + \phi)$) the RMS amplitude is $a^2/2$ at frequency f, whereas the two-side spectrum has a magnitude of $a^2/4$ at both the frequency values $-f$ and $+f$.

Note that, even though it is possible to interpret the meaning of a negative time (which represents the past, previous to the starting point), it is not possible to give a realistic meaning to a negative frequency. This concept is introduced primarily for analytical convenience.

5.2.7   Complex Spectrum

We have shown that for unique representation of a signal in the frequency domain, both amplitude and phase information should be provided for each frequency component. Alternatively, the spectrum can be expressed as a *complex* function of frequency, having a *real part* and an *imaginary part*. For instance, for a harmonic component given by $a \cos(2\pi f_i + \phi)$ the two-sided complex spectrum can be expressed as

$$Y(f_i) = \frac{a_i}{2}(\cos \phi_i + j \sin \phi_i) = \frac{a_i}{2} e^{j\phi_i}$$

and,

$$Y(-f_i) = \frac{a_i}{2}(\cos \phi_i - j \sin \phi_i) = \frac{a_i}{2} e^{-j\phi_i} \qquad (5.7)$$

in which j is the *imaginary unity* as given by $j = \sqrt{-1}$. Note that the spectral component at the negative frequency is the *complex conjugate* of that at the positive frequency. This concept of complex spectrum is the basis of (complex) *Fourier series expansion* (FSE), which we shall consider in detail in a later section. It should be clear that the complex conjugate of a spectrum is obtained by changing either j to $-$j or ω to $-\omega$ (or f to $-f$).

5.3 Signal Types

Signals can be classified into different types depending on their characteristics. Note that the signal itself is a time function, but its frequency-domain representation can bring up some of its salient features. Signals particularly important to us here are the excitations and responses of vibrating systems. These can be divided into two broad classes: *deterministic signals* and *random signals* depending on whether we are dealing with deterministic vibrations or random vibrations. Consider a damped cantilever beam that is subjected to a sinusoidal base excitation of frequency ω and amplitude u_0 in the lateral direction (Figure 5.8). In the steady state, the tip of the beam will also oscillate at the same frequency,

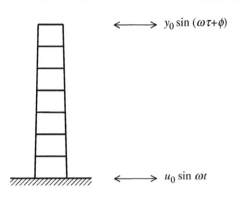

$y_0 \sin (\omega \tau + \phi)$

$u_0 \sin \omega t$

FIGURE 5.8 Response to base excitations of a tall structure (cantilever).

but with a different amplitude y_0 and, furthermore, there will be a phase shift by an angle ϕ. For a given frequency and known beam properties, the quantities y_0 and ϕ can be completely determined. Under these conditions the tip response of the cantilever is a deterministic signal in the sense that when the experiment is repeated, the same response is obtained. Furthermore, the response can be expressed as a mathematical relationship in terms of parameters whose values are determined with 100% certainty, and probabilities are not associated with these parameters (such parameters are termed *deterministic parameters*). Random signals are nondeterministic (or *stochastic*) signals. Their mathematical representation requires probability considerations. Furthermore, if the process were to be repeated there would always be some uncertainty as to whether an identical response signal could be obtained again.

Deterministic signals can be classified as *periodic*, *quasi-periodic*, and *transient*. Periodic signals repeat exactly at equal time periods. The frequency (Fourier) spectrum of a periodic signal constitutes a series of equally spaced impulses. Furthermore, a periodic signal will have a Fourier series representation. This implies that a periodic signal can be expressed as the sum of sinusoidal components whose frequency ratios are rational numbers (not necessarily integers). Quasi-periodic (or almost periodic) signals also have discrete Fourier spectra, but the spectral lines are not equally spaced. Typically, a quasi-periodic signal can be generated by combining two or more sinusoidal components, provided that at least two of the components have as their frequency ratio an irrational number. Transient signals have continuous Fourier spectra. These types of signals cannot be expressed as a sum of sinusoidal components (or a Fourier series). All signals that are not periodic or quasi-periodic can be classified as transient. Most often, highly damped (overdamped) signals with exponentially decaying characteristics are termed transient, even though various other forms of signals such as exponentially increasing (unstable) responses, sinusoidal decays (underdamped responses), and sinesweeps (sinewaves with variable frequency) also fall into this category. Table 5.1 gives examples for these three types of deterministic signals. The corresponding amplitude spectra are sketched in Figure 5.9. A general classification of signals, with some examples is given in Box 5.1.

5.4 Fourier Analysis

Fourier analysis is the key to frequency analysis of vibration signals. The frequency-domain representation of a time signal is obtained through the Fourier transform. One immediate advantage

TABLE 5.1 Deterministic Signals

Primary Classification	Nature of the Fourier Spectrum	Example
Periodic	Discrete and equally spaced	$y_0 \sin \omega t + y_1 \sin \left(\frac{5}{3} \omega t + \phi \right)$
Quasi-periodic	Discrete and irregularly spaced	$y_0 \sin \omega t + y_1 \sin(\sqrt{2}\omega t + \phi)$
Transient	Continuous	$y_0 \exp(-\lambda t)\sin(\omega t + \phi)$

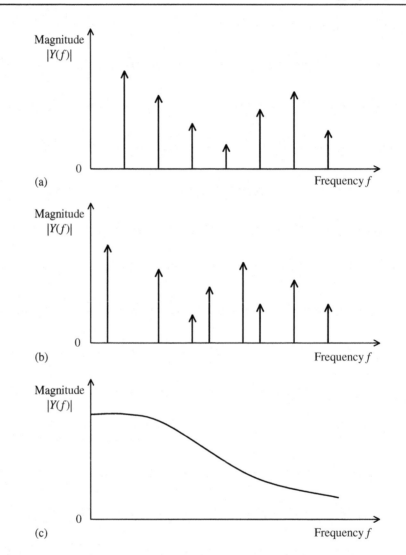

FIGURE 5.9 Magnitude spectra for three types of deterministic signals. (a) Periodic; (b) quasi-periodic; (c) transient.

of the Fourier transform is that, through its use, differential operations (differentiation and integration) in the time domain are converted into simpler algebraic operations (multiplication and division). Transform techniques are quite useful in mathematical applications. For example, a simple, yet versatile transformation from products into sums is accomplished through the use of the *logarithm*. Three versions of Fourier transform are important to us. The Fourier integral transform (FIT) can be applied to any general signal, whereas the FSE is applicable only to periodic signals, and the discrete Fourier

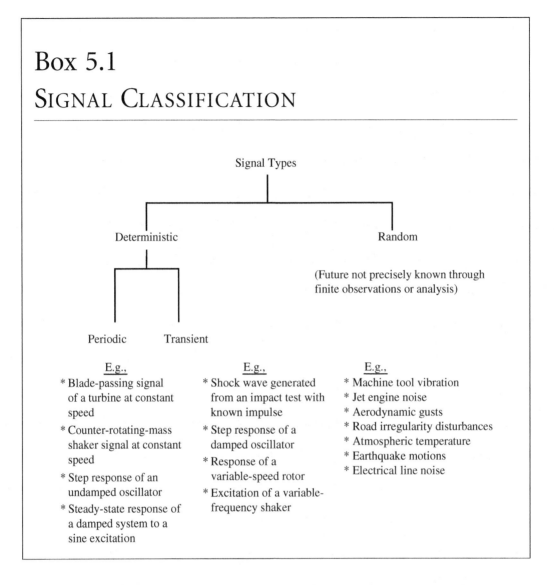

Box 5.1

SIGNAL CLASSIFICATION

Signal Types

Deterministic

Random

(Future not precisely known through
finite observations or analysis)

Periodic   Transient

E.g.,	E.g.,	E.g.,

E.g.,
* Blade-passing signal
  of a turbine at constant
  speed
* Counter-rotating-mass
  shaker signal at constant
  speed
* Step response of an
  undamped oscillator
* Steady-state response of
  a damped system to a
  sine excitation

E.g.,
* Shock wave generated
  from an impact test with
  known impulse
* Step response of a
  damped oscillator
* Response of a
  variable-speed rotor
* Excitation of a variable-
  frequency shaker

E.g.,
* Machine tool vibration
* Jet engine noise
* Aerodynamic gusts
* Road irregularity disturbances
* Atmospheric temperature
* Earthquake motions
* Electrical line noise

transform (DFT) is used for discrete signals. As we shall see, all three versions of transform are interrelated. In particular, we have to use the DFT in digital computation of both FIT and FSE.

5.4.1 Fourier Integral Transform

The *Fourier spectrum* $X(f)$ of a time signal $x(t)$ is given by the *forward transform* relation

$$X(f) = \int_{-\infty}^{\infty} x(t)\exp(-j2\pi ft)dt \qquad (5.8)$$

with $j = \sqrt{-1}$ and f the cyclic frequency variable. When Equation 5.8 is multiplied by $\exp(j2\pi f\tau)$ and integrated with respect to f using the *orthogonality property* (which can be considered as a definition of the *Dirac delta function* δ)

$$\int_{-\infty}^{\infty} \exp[j2\pi f(t - \tau)]dt = \delta(t - \tau) \qquad (5.9)$$

we get the *inverse transform* relation

$$x(t) = \int_{-\infty}^{\infty} X(f) \exp(2j\pi ft)df \qquad (5.10)$$

The forward transform is denoted by the operator $\mathfrak{F}$ and the inverse transform by $\mathfrak{F}^{-1}$. Hence, the Fourier transform pair is given by

$$X(f) = \mathfrak{F}x(t) \text{ and } x(t) = \mathfrak{F}^{-1}X(f) \qquad (5.11)$$

Note that for *real* systems, $x(t)$ is a real function but $X(f)$ is a *complex* function in general. Hence, the Fourier spectrum of a signal can be represented by the *magnitude* $|X(f)|$ and the *phase angle* $\angle X(f)$ of the (complex) Fourier spectrum $X(f)$. Alternatively, the *real part* Re $X(f)$ and the *imaginary part* Im $X(f)$ together can be used to represent the Fourier spectrum.

According to the present definition, the Fourier spectrum is defined for negative frequency values as well as positive frequencies (i.e., a *two-sided spectrum*). The *complex conjugate* of a complex value is obtained by simply reversing the sign of the imaginary part; in other words, replacing j with $-$j. By noting that replacing j with $-$j in the forward transform relation is identical to replacing f with $-f$, it should be clear that the Fourier spectrum (of real signals) for negative frequencies is given by the complex conjugate $X^*(f)$ of the Fourier spectrum for positive frequencies. As a result, only the positive-frequency spectrum needs to be specified and the negative-frequency spectrum can be conveniently derived from it, through complex conjugation.

The *Laplace transform* is similar to the FIT. Laplace transform is defined by the forward and inverse relations

$$X(s) = \int_{0}^{\infty} x(t) \exp(-st)dt \qquad (5.12)$$

and

$$x(t) = \frac{1}{2\pi j} \int_{\sigma-j\infty}^{\sigma+j\infty} X(s) \exp(st)ds \qquad (5.13)$$

Since the signal itself is zero for $t < 0$, it is seen that for all practical purposes, Fourier transform results can be deduced from the Laplace transform analysis, simply by substituting $s = j2\pi f = j\omega$ and $\sigma = 0$.

5.4.2 Fourier Series Expansion

For a periodic signal $x(t)$ of period T, the FSE is given by

$$x(t) = \Delta F \sum_{n=-\infty}^{\infty} A_n \exp(j2\pi nt/T) \qquad (5.14)$$

with $\Delta F = 1/T$. Strictly speaking (see FIT relations) this is the inverse transform relation. The scaling factor ΔF is not essential but is introduced so that the Fourier coefficients A_n will have the same units as the Fourier spectrum. The Fourier coefficients are obtained by multiplying the inverse transform relation by $\exp(-j2\pi mt/T)$ and integrating with respect to t from 0 to T using the orthogonality condition

$$\frac{1}{T} \int_{0}^{T} \exp[j2\pi(n-m)t/T]dt = \delta_{mn} \qquad (5.15)$$

Note that the *Kronecker delta* δ_{mn} is defined as

$$\delta_{mn} = \begin{cases} 1 & \text{for } m = n \\ 0 & \text{for } m \neq n \end{cases} \qquad (5.16)$$

for integer values of m and n. The forward transform that results is given by

$$A_n = \int_0^T x(t) \exp(-j2\pi nt/T) dt \qquad (5.17)$$

Note that A_n are complex quantities in general.

It can be shown that for periodic signals, FSE is a special case of FIT, as expected. Consider a Fourier spectrum consisting of a sum of equidistant impulses separated by the frequency interval $\Delta F = 1/T$:

$$X(f) = \Delta F \sum_{n=-\infty}^{\infty} A_n \delta(f - n \cdot \Delta F) \qquad (5.18)$$

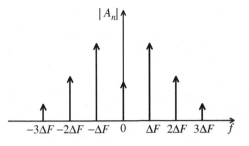

FIGURE 5.10 Fourier spectrum of a periodic signal and its relation to Fourier series.

This is shown in Figure 5.10 (only the magnitudes $|A_n|$ can be plotted in this figure because A_n is complex in general). If we substitute this spectrum into the inverse FIT relation given earlier, we get the inverse FSE relation 5.14. Furthermore, this shows that the Fourier spectrum of a periodic signal is a series of equidistant impulses.

5.4.3 Discrete Fourier Transform

The DFT relates an N-element *sequence* of *sampled* (*discrete*) data signal

$$\{x_m\} = [x_0, x_1, \ldots, x_{N-1}] \qquad (5.19)$$

to an N-element sequence of spectral results

$$\{X_n\} = [X_0, X_1, \ldots, X_{N-1}] \qquad (5.20)$$

through the forward transform relation

$$X_n = \Delta T \sum_{m=0}^{N-1} x_m \exp(-j2\pi mn/N) \qquad (5.21)$$

with $n = 0, 1, \ldots, N - 1$. The values X_n are called the *spectral lines*. It can be shown that these quantities approximate the values of the Fourier spectrum (continuous) at the corresponding discrete frequencies. Let us identify ΔT as the *sampling period* (i.e., the time step between two adjacent points of sampled data).

The inverse transform relation is obtained by multiplying the forward transform relation by $\exp(j2\pi nr/N)$ and summing over $n = 0$ to $N - 1$, using the orthogonality property

$$\frac{1}{N} \sum_{n=0}^{N-1} \exp[j2\pi n(r - m)/N] = \delta_{rm} \qquad (5.22)$$

Note that this orthogonality relation can be considered as a definition of *Kronecker delta*. The inverse transform is

$$x_m = \Delta F \sum_{n=0}^{N-1} X_n \exp(j2\pi mn/N) \qquad (5.23)$$

The data record length is given by

$$T = N \cdot \Delta T = 1/\Delta F \qquad (5.24)$$

The DFT is a transform in its own right, independent of the FIT. It is possible, however, to interpret this transformation as the *trapezoidal integration* approximation of FIT. We have deliberately chosen appropriate scaling factors ΔT and ΔF in order to maintain this equivalence, and it is very useful in computing the Fourier spectrum of a general signal or the Fourier coefficients of a periodic signal using a digital computer. Proper interpretation of the digital results is crucial, however, in using DFT to compute (an approximate) Fourier spectrum of a (continuous) signal. In particular, two types of error: aliasing

TABLE 5.2   Unified Definitions for Three Fourier Transform Types

Relation Name	Fourier Integral Transform	Discrete Fourier Transform (DFT)	Fourier Series Expansion (FSE)
Forward transform	$X(f) = \int_{-\infty}^{\infty} x(t)\exp(-j2\pi ft)\mathrm{d}t$	$X_n = \Delta T \sum_{m=0}^{N-1} x_m \exp(-j2\pi nm/N)$, $n = 0,1,\dots,N-1$	$A_n = \int_0^T x(t)\exp(-j2\pi nt/T)\mathrm{d}t$ $n = 0,1,\dots$
Inverse transform	$x(t) = \int_{-\infty}^{\infty} X(f)\exp(j2\pi ft)\mathrm{d}f$	$x_m = \Delta F \sum_{n=0}^{N-1} X_n \exp(j2\pi nm/N)$, $m = 0,1,\dots,N-1$	$x(t) = \Delta F \sum_{n=-\infty}^{\infty} A_n$ $\times \exp(j2\pi nt/T)$
Orthogonality	$\int_{-\infty}^{\infty} \exp[j2\pi f(\tau - t)]\mathrm{d}f$ $= \delta(\tau - t)$	$\frac{1}{N}\sum_{n=0}^{N-1} \exp[j2\pi n(r-m)/N] = \delta_{rm}$	$\frac{1}{T}\int_0^T \exp[j2\pi(r-n)t/T]\mathrm{d}t = \delta_{rn}$
Notes	$T = N\Delta T$	$\Delta F = 1/T$	$X(f) = \Delta F \sum_{n=-\infty}^{\infty} A_n \delta(f - n/T)$

TABLE 5.3   Important Properties of the Fourier Transform

Function of Time	Fourier Spectrum
$x(t)$	$X(f)$
$k_1 x_1(t) + k_2 x_2(t)$	$k_1 X_1(f) + k_2 X_2(f)$
$x(t)\exp(-j2\pi ta)$	$X(f + a)$
$x(t + \tau)$	$X(f)\exp(j2\pi f\tau)$
$\dfrac{\mathrm{d}^n x(t)}{\mathrm{d}t^n}$	$(j2\pi f)^n X(f)$
$\int_{-\infty}^{t} x(t)\mathrm{d}t$	$\dfrac{X(f)}{j2\pi f}$

and *leakage* (or *truncation error*) should be considered. This subject will be addressed later. The three transform relations, corresponding inverse transforms, and the orthogonality relations are summarized in Table 5.2.

The link between the time-domain signals and models and the corresponding frequency-domain equivalents is the FIT. Table 5.3 provides some important properties of the FIT and the corresponding time-domain relations that are useful in the analysis of signals and system models. These properties may be easily derived from the basic FIT relations (Equation 5.8 through Equation 5.10). It should be noted that, inherent in the definition of the DFT given in Table 5.2 is the N-point periodicity of the two sequences; that is, $X_n = X_{n+iN}$ and $x_m = x_{m+iN}$, for $i = \pm1, \pm2, \dots$.

The definitions given in Table 5.2 may differ from the versions available in the literature by a multiplicative constant. However, it is observed that according to the present definitions, the DFT may be interpreted as the trapezoidal integration of the FIT. The close similarity between the definitions of the FSE and the DFT is also noteworthy. Furthermore, according to the last row in Table 5.2, the FSE can be expressed as a special FIT consisting of an equidistant set of impulses of magnitude A_n/T located at $f = n/T$.

5.4.4   Aliasing Distortion

Recalling that the primary task of digital Fourier analysis is to obtain a discrete approximation to the FIT of a piecewise continuous function, it is advantageous to interpret the DFT as a discrete (digital computer) version of the FIT rather than an independent discrete transform. Accordingly, the results from a DFT must be consistent with the exact results obtained if the FIT were used. The definitions given in Table 5.2 are consistent in this respect because the DFT is given as the trapezoidal integration of the FIT. However, it should be clear that if $X(f)$ is the FIT of $x(t)$, then the sequence of sampled values $\{X(n \cdot \Delta F)\}$ is not exactly the DFT of the sampled data sequence $\{x(m \cdot \Delta T)\}$. Only an approximate relationship exists.

A further advantage of the definitions given in Table 5.2 is apparent when dealing with the FSE. As we have noted, the FIT of a periodic function is a set of impulses. We can avoid dealing with impulses by relating the complex Fourier coefficients to the DFT sequence of sampled data from the periodic function via the present definitions.

Aliasing distortion is an important consideration when dealing with sampled data from a continuous signal. This error may enter into computation in both the time domain and the frequency domain, depending on the domain in which the results are presented. We will address this issue next.

5.4.4.1 Sampling Theorem

The basic relationships between the FIT, the DFT, and the FSE are summarized in Table 5.4. By means of straightforward mathematical procedures, the relationship between the FIT and the DFT can be established. Even though $\{X(n \cdot \Delta F)\}$ is not the DFT of $\{x(m \cdot \Delta T)\}$, the results in Table 5.4 show that $\{\tilde{X}(n \cdot \Delta F)\}$ is the DFT of $\{\tilde{x}(m \cdot \Delta T)\}$ where the periodic functions $\tilde{X}(f)$ and $\tilde{x}(t)$ are as defined as in Table 5.4. This situation is illustrated in Figure 5.11. It should be recalled that $X(f)$ is a complex function in general, and as such it cannot be displayed as a single curve in a two-dimensional coordinate system. Both the magnitude and the phase angle variations with respect to frequency f are needed. For brevity, only the magnitude $|X(f)|$ is shown in Figure 5.11(a). Nevertheless, the argument presented applies to the phase angle $\angle X(f)$ as well.

It is obvious that in the time interval $[0, T]$, $x(t) = \tilde{x}(t)$ and $x_m = \tilde{x}_m$. However, $\tilde{X}(n \cdot \Delta F)$ is only approximately equal to $X(n \cdot \Delta F)$ in the frequency interval $[0, F]$. This is known as the aliasing distortion in the frequency domain. As ΔT decreases (i.e., as F increases) $\tilde{X}(f)$ will become closer to $X(f)$ in the frequency interval $[0, F/2]$, as is clear from Figure 5.11(c). Furthermore, due to the F-periodicity of $\tilde{X}(f)$, its value in the frequency range $[F/2, F]$ will approximate $X(f)$ in the frequency range $[-F/2, 0]$.

It is clear from the preceding discussion that if a time signal $x(t)$ is sampled at equal steps of ΔT, no information regarding its frequency spectrum $X(f)$ is obtained for frequencies higher than $f_c = 1/(2\Delta T)$. This fact is known as *Shannon's sampling theorem*, and the limiting (cut-off) frequency is called the *Nyquist frequency*. In vibration signal analysis, a sufficiently small sample step ΔT should be chosen in order to reduce aliasing distortion in the frequency domain, depending on the highest frequency of interest in the analyzed signal. This however, increases the signal processing time and the computer storage requirements, which is undesirable, particularly in real-time analysis. It can also result in stability problems in numerical computations. The Nyquist sampling criterion requires that the sampling rate $(1/\Delta T)$ for a signal should be at least twice the highest frequency of interest. Instead of making the sampling rate very high, a moderate value that satisfies the Nyquist sampling criterion is used in practice, together with an *antialiasing filter* to remove the distorted frequency components. It should be noted that the DFT results in the frequency interval $[f_c, 2f_c]$ are redundant because they merely approximate the frequency spectrum in the negative frequency interval $[-f_c, 0]$ which is known for real signals. This fact is known as the Hermitian property.

The last column of Table 5.4 presents the relationship between the FSE and the DFT. It is noted that the sequence $\{\tilde{A}_n\}$ rather than the sequence of complex Fourier series coefficients $\{A_n\}$ represents the DFT of

TABLE 5.4 Unified Fourier Transform Relationships

Description	Relationship	
	DFT and FIT	DFT and FSE
Given	$x(t) \overset{\text{FIT}}{\to} X(f)$	$x(t) \overset{\text{FSE}}{\to} \{A_n\}$
Form	$\tilde{x}(t) = \sum_{k=-\infty}^{\infty} x(t + kT)$	$\tilde{A}_n = \sum_{k=-\infty}^{\infty} A_{n+kN}$
	$\tilde{X}(f) = \sum_{k=-\infty}^{\infty} X(f + kF)$	
Then	$\{\tilde{x}_m\} \overset{\text{DFT}}{\to} \{\tilde{X}_n\}$	$\{x_m\} \overset{\text{DFT}}{\to} \{\tilde{A}_n\}$
Where	$\tilde{x}_m = \tilde{x}(m \cdot \Delta T), \tilde{X}_n = \tilde{X}(n \cdot \Delta F)$	$x_m = x(m \cdot \Delta T)$
	$F = 1/\Delta T, T = 1/\Delta F$	$N = T/\Delta T$

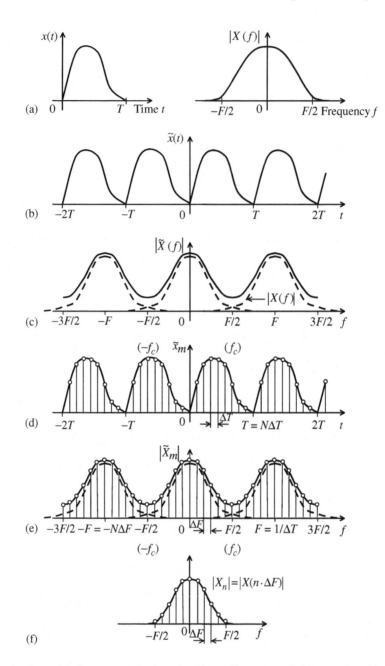

FIGURE 5.11 Relationship between FIT and DFT, with an illustration of aliasing error. (a) Fourier integral transformation (FIT) of a signal; (b) periodically arranged time signal; (c) periodically overlapped frequency spectrum; (d) sampled time signal (b); (e) sampled frequency spectrum (c) (with aliasing error); (f) sampled original spectrum (no aliasing error).

the sampled data sequence $\{x(m \cdot \Delta T)\}$. In practice, however, $A_n \rightarrow 0$ as $n \rightarrow \infty$. Consequently, $\tilde{A}_n$ is a good approximator to A_n in the range $[-N/2 \leq n \leq N/2]$ for sufficiently large N. This basic result is useful in determining the Fourier coefficients of a periodic signal using discrete data that are sampled at time steps of $\Delta T = 1/F$, in which F is the fundamental frequency of the periodic signal. Again the aliasing error $(\tilde{A}_n - A_n)$ may be reduced by increasing the sampling rate (i.e., by decreasing ΔT or increasing N).

5.4.4.2 Aliasing Distortion in the Time Domain

In vibration applications it is sometimes required to reconstruct the signal from its Fourier spectrum. Inverse DFT is used for this purpose and is particularly applicable in digital equalizers in vibration testing. Due to sampling in the frequency domain, the signal becomes distorted. The aliasing error $(\tilde{x}_m - x(m\Delta T))$ is reduced by decreasing the sample period ΔF. It should be noted that no information regarding the signal for times greater than $T = 1/\Delta F$ is obtained from the analysis.

By comparing Figure 5.11(a) with (c), or (e) with (f), it should be clear that the aliasing error in $\tilde{X}$ in comparison with the original spectrum X is caused by "folding" of the high-frequency segment of X beyond the Nyquist frequency into the low-frequency segment of X. This is illustrated in Figure 5.12.

5.4.4.3 Antialiasing Filter

It should be clear from Figure 5.12 that, if the original signal is low-pass filtered at a cut-off frequency equal to the Nyquist frequency, then the aliasing distortion would not occur due to sampling. A filter of this type is called an antialiasing filter. In practice, it is not possible to achieve perfect filtering. Hence, some aliasing could remain even after using an antialiasing filter. Such residual errors may be reduced by using a filter cut-off frequency that is slightly less than the Nyquist frequency. The resulting spectrum would then only be valid up to this filter cut-off frequency (and not up to the theoretical limit of Nyquist frequency).

Example 5.1

Consider 1024 data points from a signal, sampled at 1 msec intervals.

$$\text{Sample rate } f_s = 1/0.001 \text{ samples/sec} = 1000 \text{ Hz} = 1 \text{ kHz}$$

$$\text{Nyquist frequency} = 1000/2 \text{ Hz} = 500 \text{ Hz}$$

Due to aliasing, approximately 20% of the spectrum (i.e., spectrum beyond 400 Hz) will be distorted. Here we may use an antialiasing filter.

Suppose that a digital Fourier transform computation provides 1024 frequency points of data up to

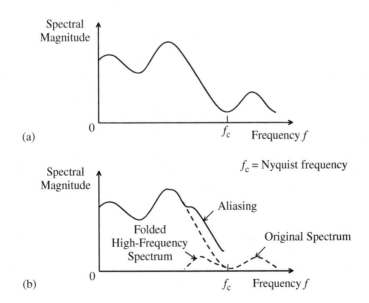

FIGURE 5.12 Aliasing distortion of frequency spectrum. (a) Original spectrum; (b) distorted spectrum due to aliasing.

Box 5.2

SIGNAL SAMPLING CONSIDERATIONS

The maximum useful frequency in digital Fourier results is half the sampling rate.
Nyquist frequency or cut-off frequency or computational bandwidth:

$$f_c = \frac{1}{2} \times \text{sampling rate}$$

Aliasing distortion:
High-frequency spectrum beyond Nyquist frequency folds on to the useful spectrum, thereby distorting it.
Summary:

1. Pick a sufficiently small sample step ΔT in the time domain, to reduce the aliasing distortion in the frequency domain.
2. The highest frequency for which the Fourier transform (frequency-spectrum) information would be valid, is the Nyquist frequency $f_c = 1/(2\Delta T)$.
3. DFT results that are computed for the frequency range $[f_c, 2f_c]$ merely approximate the frequency spectrum in the negative frequency range $[-f_c, 0]$.

1000 Hz. Half of this number is beyond the Nyquist frequency and will not give any new information about the signal.

$$\text{Spectral line separation} = 1000/1024 \text{ Hz} = 1 \text{ Hz (approximately)}$$

Keep only the first 400 spectral lines as the useful spectrum.
Note: Almost 500 spectral lines may be retained if an accurate antialiasing filter is used.
Some useful information on signal sampling is summarized in Box 5.2.

5.4.5   Another Illustration of Aliasing

A simple illustration of aliasing is given in Figure 5.13. Here, two sinusoidal signals of frequency, $f_1 = 0.2$ Hz and $f_2 = 0.8$ Hz, are shown (Figure 5.13(a)). Suppose that the two signals are sampled at the rate of $f_s = 1$ sample/sec. The corresponding Nyquist frequency is $f_c = 0.5$ Hz. It is seen that, at this sampling rate, the data samples from the two signals are identical. In other words, the high-frequency signal cannot be distinguished from the low-frequency signal. Hence, a high-frequency signal component of frequency 0.8 Hz will appear as a low-frequency signal component of frequency 0.2 Hz. This is aliasing, as is clear from the signal spectrum shown in Figure 5.13. Specifically, the spectral segment of the signal beyond the Nyquist frequency (f_c) cannot be recovered.

It is apparent from Figure 5.11(e) that the aliasing error becomes increasingly prominent for frequencies of the spectrum closer to the Nyquist frequency. With reference to the expression for $\tilde{X}(f)$ in Table 5.4, it should be clear that when the true Fourier spectrum $X(f)$ has a steep roll-off prior to $F/2$ ($= f_c$), the influence of the $X(f - nF)$ segments for $n \geq 2$ and $n \leq -1$ is negligible in the discrete spectrum in the frequency range $[0, F/2]$. Hence the aliasing distortion in the frequency band $[0, F/2]$ comes primarily from $X(f - F)$, which is the true spectrum shifted to the right through F. Therefore,

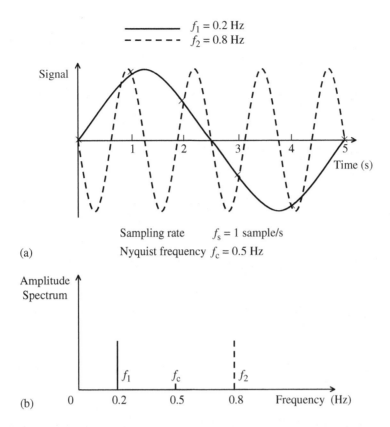

FIGURE 5.13 A simple illustration of aliasing. (a) Two harmonic signals with identical sampled data; (b) Frequency spectra of the two harmonic signals.

a reasonably accurate expression for the aliasing error is

$$e_n = X(-(F - n \cdot \Delta F)) = X(n \cdot \Delta F - F) = X_{-(N-n)} = X_{n-N} \qquad n = 0, 1, 2, ..., N/2 \qquad (5.25)$$

Note from Equation 5.8 that the spectral value obtained when f in the complex exponential is replaced by $-f$ is the same as the spectral value obtained when j is replaced by $-j$. Since the signal $x(t)$ is real, it follows that the Fourier spectrum for the negative frequencies is simply the complex conjugate of the Fourier spectrum for the positive frequencies; thus

$$X(-f) = X^*(f) \qquad (5.26)$$

or, in the discrete case

$$X_{n-N} = X^*_{N-n} \qquad (5.27)$$

It follows from Equation 5.25 that the aliasing distortion is given by

$$e_n = X^*_{N-n} \qquad \text{for } n = 0, 1, 2, ..., N/2 \qquad (5.28)$$

This result confirms that aliasing can be interpreted as folding of the complex conjugate of the true spectrum beyond the Nyquist frequency $f_c (= F/2)$ over to the original spectrum. In other words, due to aliasing, frequency components higher than the Nyquist frequency appear as lower frequency components (due to folding). These aliasing components enter into the digital Fourier results in the useful frequency range $[0, f_c]$.

Aliasing reduces the valid frequency range in digital Fourier results. Typically, the useful frequency limit is $f_c/1.28$ so that the last 20% of the spectral points near the Nyquist frequency should be neglected.

It should be clear that if a low-pass filter with its cut-off frequency set at f_c is used on the time signal prior to sampling and digital Fourier analysis, the aliasing distortion can be virtually eliminated. Analog hardware filters may be used for this purpose. They are the *antialiasing filters*. Note that sometimes $f_c/1.28 (\cong 0.8 f_c)$ is used as the filter cut-off frequency. In this case, the computed spectrum is accurate up to $0.8 f_c$ and not up to f_c.

The buffer memory of a typical commercial Fourier analyzer can store $N = 2^{10} = 1024$ samples of data from the time signal. This is the size of the data block analyzed in each digital Fourier transform calculation. This will result in $N/2 = 512$ spectral points (spectral lines) in the frequency range $[0, f_c]$. Out of this only the first 400 spectral lines (approximately 80%) are considered free of aliasing distortion.

Example 5.2

Suppose that the frequency range of interest in a particular vibration signal is 0–200 Hz. We are interested in determining the sampling rate (digitization speed) and the cut-off frequency for the antialiasing (low-pass) filter.

$$\text{The Nyquist frequency } f_c \text{ is given by } f_c/1.28 = 200 \text{ Hz}$$

$$\text{Hence, } f_c = 256 \text{ Hz}$$

The sampling rate (or digitization speed) for the time signal that is needed to achieve this range of analysis is $F = 2 f_c = 512$ Hz. With this sampling frequency, the cut-off frequency for the antialiasing filter could be set at a value between 200 and 256 Hz.

5.5 Analysis of Random Signals

Random (stochastic) signals are generated by some random mechanism. Each time the mechanism is operated a new time history (sample function) is generated. The chance of any two sample functions becoming identical is governed by some probabilistic law. If all sample functions are identical (with unity probability), then the corresponding signal is a deterministic signal. A random process is denoted by $\tilde{X}(t)$, while any sample function of it is denoted by $x(t)$. No numerical computations can be performed on $\tilde{X}(t)$ because it is not known for certain. Its Fourier transform, for instance, can be written down as an analytical expression, but cannot be numerically computed. However, once the signal is generated, numerical computations can be performed on that sample function $x(t)$ because it is a completely known function of time.

5.5.1 Ergodic Random Signals

At any given time t_1, $\tilde{X}(t_1)$ is a random variable which has a certain probability distribution. Consider a well-behaved function $f\{\tilde{X}(t)\}$ of this random variable (which is also a random variable). Its expected value (statistical mean) is $E[f\{\tilde{X}(t)\}]$. This is also known as the ensemble average because it is equivalent to the average value at t of a collection (ensemble) of a large number of sample functions $x(t)$.

Consider the function $f\{x(t)\}$ of *one* sample function $x(t)$. Its temporal (time) mean is expressed by

$$\lim_{T \to \infty} \frac{1}{2T} \int_{-T}^{T} f\{x(t)\} dt$$

Now, if

$$E[f\{\tilde{X}(t_1)\}] = \lim_{T \to \infty} \frac{1}{2T} \int_{-T}^{T} f\{x(t)\} dt \tag{5.29}$$

then the random signal is said to be ergodic. It should be noted that the right-hand side of Equation 5.29 does not depend on time. Consequently, the left-hand side should also be independent of the time point t_1.

For analytical convenience, random vibration signals are usually assumed to be ergodic (the ergodic hypothesis). Using this hypothesis, the properties of a random signal could be determined by performing computations on a sufficiently long record (sample function) of the signal. Since the ergodic hypothesis is not exactly satisfied for vibration signals, and since it is impossible to analyze infinitely long data records, the accuracy of the numerical results depends on various factors such as the record length, sampling rate, frequency range of interest, and the statistical nature of the random signal (e.g., closeness to a deterministic signal, frequency content, periodicity, damping characteristics). Accuracy can be improved in general, by averaging the results for more than one data record.

5.5.2 Correlation and Spectral Density

If for a random signal $\tilde{X}(t)$, the joint statistical properties of $\tilde{X}(t_1)$ and $\tilde{X}(t_2)$ depend on the time difference $(t_2 - t_1)$ and not on t_1 itself, then the signal is said to be stationary. Consequently, the statistical properties of a stationary $\tilde{X}(t)$ will be independent of t. It is noted from Equation 5.29 that ergodic random signals are necessarily stationary. However, in general the converse is not true.

The cross-correlation function of two random signals $\tilde{X}(t)$ and $\tilde{Y}(t)$ is given by $E[\tilde{X}(t)\tilde{Y}(t + \tau)]$. If the signals are stationary, this expected value is a function of τ (not t) and is denoted by $\phi_{xy}(\tau)$. In view of the ergodic hypothesis, the cross-correlation function may be expressed as

$$\phi_{xy}(\tau) = \lim_{T \to \infty} \left[\frac{1}{T} \int_0^T x(t)y(t + \tau)dt \right] \tag{5.30}$$

The FIT of $\phi_{xy}(\tau)$ is the cross-spectral density function which is denoted by $\Phi_{xy}(f)$. When the two signals are identical, we have the autocorrelation function $\phi_{xx}(\tau)$ in the time domain and the power spectral density (PSD) $\Phi_{xx}(f)$ in the frequency domain. The continuous and the discrete versions of the correlation theorem are given in the first row of Table 5.5. It follows that the cross-spectral density may be estimated using the DFT (FFT) of the two signals, as $[X_n]^* Y_n / T$ in which T is the record length and $[X_n]^*$ is the complex conjugate of $[X_n]$.

Parseval's theorem (second row of Table 5.5) follows directly from the correlation theorem. Consequently, the mean square value of a random signal may be obtained from the area under the PSD curve. This suggests a hardware-based method of estimating the PSD as illustrated by the functional diagram in Figure 5.14(a). Alternatively, a software-based digital Fourier analysis could be used (Figure 5.14(b)). A single sample function would not give the required accuracy, and averaging is usually needed. In real-time digital analysis, the running average as well as the current estimate is usually computed. In the running average, it is desirable to give a higher weighting to the more recent estimates. The fluctuations in the PSD estimate about the local average could be reduced by selecting a large filter

TABLE 5.5 Some Useful Fourier Transform Results

Description	Continuous	Discrete
Correlation theorem	If $z(\tau) = \int_{-\infty}^{\infty} x(t)y(t + \tau)dt$	$z_m = \Delta T \sum_{r=0}^{N-1} x_r y_{r+m}$
	Then $Z(f) = [X(f)]^* Y(f)$	$Z_n = [X_n]^* Y_n$
Parseval's theorem	If $y(t) \overset{\text{FIT}}{\to} Y(f)$	$\{y_m\} \overset{\text{DFT}}{\to} \{Y_n\}$
	Then $\int_{-\infty}^{\infty} y^2(t)dt = \int_{-\infty}^{\infty} |Y(f)|^2 df$	$\Delta T \sum_{m=0}^{N-1} y_m^2 = \Delta F \sum_{n=0}^{N-1} |Y_n|^2$
Convolution theorem	If $y(t) = \int_{-\infty}^{\infty} h(\tau)u(t - \tau)d\tau = \int_{-\infty}^{\infty} h(t - \tau)u(\tau)d\tau$	$y_m = \Delta T \sum_{r=0}^{N-1} h_r u_{m-r} = \Delta T \sum_{r=0}^{N-1} h_{m-r} u_r$
	Then $Y(f) = H(f)U(f)$	$Y_n = H_n U_n$

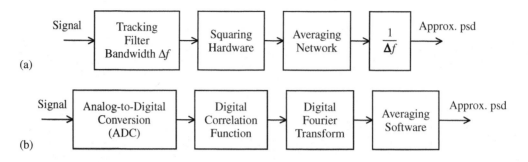

FIGURE 5.14 Power spectral density computation. (a) Narrow-band filtering method; (b) correlation and Fourier transformation method.

bandwidth Δf and a large record length T. A measure of these fluctuations is given by

$$\varepsilon = \frac{1}{\sqrt{\Delta f T}} \qquad (5.31)$$

It should be noted that a large Δf results in reduction of the precision of the estimates while improving the appearance. To offset this, T has to be increased further.

5.5.3 Frequency Response Using Digital Fourier Transform

Vibration test programs usually require a resonance search type pretesting. In order to minimize the damage potential, it is carried out at a much lower intensity than the main test. The objective of such exploratory tests is to determine the significant frequency-response functions of the test specimen. These provide the natural frequencies, damping ratios, and mode shapes of the test specimen. Such frequency-response data are useful in planning and conducting the main test. For example, more attention is required when testing in the vicinity of the resonance points (slower sweep rates, larger dwell periods, etc.). Also, the frequency-response data are useful in determining the most desirable test input directions and intensities. The degree of nonlinearity and time variance of the test object can be determined by conducting more than one frequency-response test at different input intensities. If the deviation of the frequency-response function is sufficiently small, then linear, time-invariant analysis is considered to be satisfactory. Often, frequency-response tests are conducted at full test intensity. In such cases, it is considered as a part of the main test rather than a prescreening test. Other uses of the frequency-response function include the following: it can be employed as a system model (experimental model) for further analysis of the test specimen (experimental modal analysis). Most desirable frequency range and sweep rates for vibration testing can be estimated by examining frequency-response functions.

The time response $h(t)$ to a unit impulse is known as the impulse-response function. For each pair of input and output locations (A,B) of the test specimen, a corresponding single response function would be obtained (assuming linearity and time-invariance). Entire collection of these responses would determine the response of the test specimen to an arbitrary input signal. The response $y(t)$ at B to an arbitrary input $u(t)$ applied at A, is given by

$$y(t) = \int_{-\infty}^{\infty} h(\tau)u(t - \tau)\mathrm{d}\tau = \int_{-\infty}^{\infty} h(t - \tau)u(\tau)\mathrm{d}\tau \qquad (5.32)$$

The right-hand side of Equation 5.32 is the convolution integral of $h(t)$ and $u(t)$ and is denoted by $h(t) * u(t)$. By substituting the inverse FIT relations (Table 5.2) in Equation 5.32, the frequency-response function (frequency-transfer function) $H(f)$ is obtained as the ratio of the (complex) FITs of the output and the input. It exists for physically realizable (casual) systems even when the individual FITs of the input and output signals do not converge. The continuous convolution theorem and the discrete counterpart are given in the last row of Table 5.5. The discrete convolution can be interpreted as the trapezoidal integration of Equation 5.32. Frequency-response function is a valid representation (model) for linear, time-invariant systems. It is related to the system transfer function $G(s)$ (ratio of the Laplace transforms of output and input with zero initial conditions) through

$$H(f) = \frac{Y(f)}{U(f)} = G(2\pi f j) \tag{5.33}$$

But for notational convenience, the frequency-response function corresponding to $G(s)$ may be denoted by either $G(f)$ or $G(\omega)$, where ω is the angular frequency and f is the cyclic frequency.

Using Fourier transform theory, three methods of determining $H(f)$ can be established. First, using any transient excitation signal to a system at rest and the corresponding output, $H(f)$ is determined from their FITs (Table 5.5). Second, if the input is sinusoidal, the signal amplification of the steady-state output is the magnitude $|H(f)|$ at the input frequency, and the phase lead of the steady-state output is the corresponding phase angle $\angle H(f)$. Third, using a random input signal and the corresponding input and output spectral density functions, $H(f)$ is determined as the ratio

$$H(f) = \Phi_{uy}(f)/\Phi_{uu}(f) \tag{5.34}$$

5.5.4 Leakage (Truncation Error)

In digital processing of vibration signals (e.g., accelerometer signals), sampled data are truncated to eliminate less significant parts. This is of course essential in real-time processing because, in that case, only sufficiently short segments of continuously acquisitioned data are processed at a time. The computer memory (and buffer) limitations, the speed and cost of processing, the frequency range of importance, the sampling rate, and the nature of the signal (level of randomness, periodicity, decay rate, etc.) should be taken into consideration in selecting the truncation point of data.

The effect of direct truncation of a signal $x(t)$ on its Fourier spectrum is shown in Figure 5.15. In the time domain, truncation is accomplished by multiplying $x(t)$ by the box-car function $b(t)$. This is equivalent to a convolution $(X(f) * B(f))$ in the frequency domain. This procedure introduces ripples (side lobes) into the true spectrum. The resulting error $(X(f) - X(f) * B(f))$ is known as leakage or truncation error. Similar leakage effects arise in the time domain, as a result of truncation of the frequency spectrum. The truncation error may be reduced by suppressing the side lobes, which requires modification of the truncation function (window) from the box-car shape $b(t)$ to a more desirable shape. Commonly used windows are the Hanning, Hamming, Parzen, and Gaussian windows.

5.5.5 Coherence

Random vibration signals $\tilde{X}(t)$ and $\tilde{Y}(t)$ are said to be statistically independent if their joint probability distribution is given by the product of the individual distributions. A special case of this is the uncorrelated signals which satisfy

$$E[\tilde{X}(t_1)\tilde{Y}(t_2)] = E[\tilde{X}(t_1)]E[\tilde{Y}(t_2)] \tag{5.35}$$

In the stationary case, the means $\mu_x = E[\tilde{X}(t)]$ and $\mu_y = E[\tilde{Y}(t)]$ are time independent. The autocovariance functions are given by

$$\psi_{xx}(\tau) = E[\{\tilde{X}(t) - \mu_x\}\{\tilde{X}(t+\tau) - \mu_x\}] = \phi_{xx}(\tau) - \mu_x^2 \tag{5.36}$$

$$\psi_{yy}(\tau) = E[\{\tilde{Y}(t) - \mu_y\}\{\tilde{Y}(t+\tau) - \mu_x\}] = \phi_{yy}(\tau) - \mu_y^2 \tag{5.37}$$

and the cross-covariance function is given by

$$\psi_{xy}(\tau) = E[\{\tilde{X}(t) - \mu_x\}\{\tilde{Y}(t+\tau) - \mu_y\}] = \phi_{xy}(\tau) - \mu_x\mu_y \tag{5.38}$$

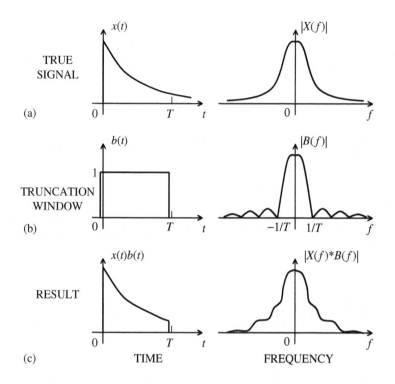

FIGURE 5.15 Illustration of truncation error. (a) Signal and its frequency spectrum; (b) a rectangular (box-car) window and its frequency spectrum; (c) truncated signal and its frequency spectrum.

For uncorrelated signals $\phi_{xy}(\tau) = \mu_x\mu_y$ and $\psi_{xy}(\tau) = 0$. The correlation function coefficient is defined by

$$\rho_{xy}(\tau) = \frac{\psi_{xy}(\tau)}{\sqrt{\psi_{xx}(0)\psi_{yy}(0)}} \tag{5.39}$$

which satisfies $-1 \leq \rho_{xy}(\tau) \leq 1$.

For uncorrelated signals we have $\rho_{xy}(\tau) = 0$. This function measures the degree of correlation of the two signals. In the frequency domain, the correlation is determined by its (ordinary) coherence function

$$\gamma_{xy}^2(f) = \frac{|\Phi_{xy}(f)|^2}{\Phi_{xx}(f)\Phi_{yy}(f)} \tag{5.40}$$

which satisfies the condition $0 \leq \gamma_{xy}^2(f) \leq 1$. In this definition, the signals are assumed to have zero means. Alternatively, the FIT of the covariance functions may be used. If the signals are uncorrelated (or better, independent) the coherence function vanishes. On the other hand, if $\tilde{Y}(t)$ is the response of a linear, time-invariant system to an input $\tilde{X}(t)$, then

$$\Phi_{xy}(f) = \Phi_{xx}(f)H(f) \tag{5.41}$$

$$\Phi_{yy}(f) = \Phi_{xx}(f)|H(f)|^2 \tag{5.42}$$

Consequently, the coherence function becomes unity for this ideal case. In practice, however, the coherence function of an excitation and the corresponding response is usually less than unity. This is due to deviations such as measurement noise, system nonlinearities, and time-variant effects. Consequently, the coherence function is commonly used as a measure of the accuracy of frequency-response estimates.

5.5.6 Parseval's Theorem

For a pair of rapidly decaying (aperiodic) deterministic signals $x(t)$ and $y(t)$, the cross-correlation function is given by

$$\phi_{xy}(\tau) = \int_{-\infty}^{\infty} x(t)y(t + \tau)dt \tag{5.43}$$

This is equivalent to Equation 5.30, for a pair of ergodic, random (stochastic) signals $x(t)$ and $y(t)$. By using the definition of the inverse FIT (see Table 5.2) in Equation 5.43, and by following straightforward mathematical manipulation, it may be shown that

$$\Phi_{xy}(f) = [X(f)]^* Y(f) \tag{5.44}$$

in which the cross-spectral density $\Phi_{xy}(f)$ is the FIT of $\phi_{xy}(\tau)$, as given by

$$\Phi_{xy}(f) = \int_{-\infty}^{\infty} \phi_{xy}(\tau) \exp(-j2\pi f\tau)d\tau \tag{5.45}$$

and $[\ ]^*$ denotes the complex conjugation operation. This result, which is known as the correlation theorem (see Table 5.5) has applications in the evaluation of the correlation functions and PSD functions of finite-record-length data.

The inverse FIT relation corresponding to Equation 5.45 is

$$\phi_{xy}(\tau) = \int_{-\infty}^{\infty} \Phi_{xy}(f) \exp(j2\pi f\tau)df \tag{5.46}$$

From Equation 5.44, we have

$$\phi_{xy}(\tau) = \int_{-\infty}^{\infty} [X(f)]^* Y(f) \exp(j2\pi f\tau)df \tag{5.47}$$

If we set $\tau = 0$ and $x = y$ in Equation 5.47, we get

$$\phi_{yy}(0) = \int_{-\infty}^{\infty} |Y(f)|^2 df \tag{5.48}$$

Similarly, from Equation 5.43, we get

$$\phi_{yy}(0) = \int_{-\infty}^{\infty} y^2(t)dt \tag{5.49}$$

By comparing Equation 5.48 and Equation 5.49, we obtain Parseval's theorem:

$$\int_{-\infty}^{\infty} y^2(t)dt = \int_{-\infty}^{\infty} |Y(f)|^2 df \tag{5.50}$$

By using the discrete correlation theorem is an analogous manner, we can establish the discrete version of Equation 5.50:

$$\Delta T \sum_{m=0}^{N-1} y_m^2 = \Delta F \sum_{n=0}^{N-1} |Y_n|^2 \tag{5.51}$$

These results are listed in the second row of Table 5.5.

5.5.7 Window Functions

Consider the unit box-car function $w(t)$, defined as

$$w(t) = \begin{cases} 1 & \text{for } -0 \le t < T \\ 0 & \text{otherwise} \end{cases} \tag{5.52}$$

TABLE 5.6 Some Common Window Functions

Function Name	Time-Domain Representation [$w(t)$]	Frequency-Domain Representation [$W(f)$]
Box-car	$\begin{cases} 1 & \text{for } 0 \leq t < T \\ 0 & \text{otherwise} \end{cases}$	$\dfrac{1}{j2\pi f}[1 - \cos 2\pi fT + j\sin 2\pi fT]$
Hanning	$\begin{cases} \dfrac{1}{2} + \dfrac{1}{2}\cos\dfrac{\pi t}{T} & \text{for } |t| < T \\ 0 & \text{otherwise} \end{cases}$	$\dfrac{T\sin 2\pi fT}{2\pi fT[1 - (2fT)^2]}$
Parzen	$\begin{cases} 1 - 6\left[\dfrac{t}{T}\right]^2 + 6\left[\dfrac{|t|}{T}\right]^3 & \text{for } |t| < \dfrac{T}{2} \\ 2\left[1 - \dfrac{|t|}{T}\right]^3 & \text{for } \dfrac{T}{2} < |t| \leq T \\ 0 & \text{otherwise} \end{cases}$	$\dfrac{3}{4}T\left[\dfrac{\sin \pi fT/2}{\dfrac{1}{2}\pi fT}\right]^4$
Bartlett	$\begin{cases} 1 - \dfrac{|t|}{T} & \text{for } |t| \leq T \\ 0 & \text{otherwise} \end{cases}$	$T\left[\dfrac{\sin \pi fT}{\pi fT}\right]^4$

This is shown in Figure 5.15(b). The FIT of $w(t)$ is

$$W(f) = \frac{1}{j2\pi f}[1 - \cos 2\pi fT + j\sin 2\pi fT] \tag{5.53}$$

Clearly, this (rectangular) window function produces side lobes (leakage) in the frequency domain.

In spectral analysis of vibration signals, it is often required to segment the time history into several parts, and then perform spectral analysis on the individual results to observe the time development of the spectrum. If segmenting is done by simple truncation (multiplication by the box-car window), the process would introduce rapidly fluctuating side lobes into spectral results. Window functions, or smoothing functions other than the box-car function, are widely used to suppress the side lobes (leakage error). Some common smoothing functions are defined in Table 5.6.

A graphical comparison of these four window types is given in Figure 5.16. Hanning windows are very popular in practical applications. A related window is the Hamming window, which is simply a Hanning window with rectangular cut-offs at the two ends. A Hamming window will have characteristics similar to those of a Hanning window, except that the side lobe fall-off rate at higher frequencies is less in the Hamming window.

From Figure 5.16(b), we observe that the frequency-domain weight of each window varies with the frequency range of interest. Obviously, the box-car window is the worst. In practical applications, the performance of any window could be improved by simply increasing the window length T. In addition, a window in the shape of a Gaussian function may be used when a rapid roll-off without side lobes is desired.

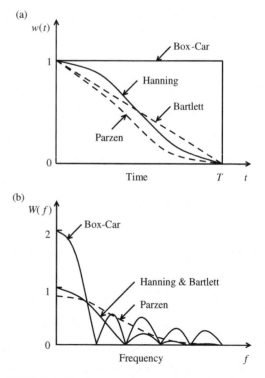

FIGURE 5.16 Some common window functions. (a) Time-domain function; (b) frequency spectrum.

TABLE 5.7 Signal Types and Appropriate Windows

Signal Type	Window
Periodic with period $= T$	Rectangular
Rapid transients within $[0, T]$	Gaussian
Periodic with period $\neq T$	Flat-top cosine
Quasi-periodic	Hamming
Slow transients beyond $[0, T]$	Hanning
Nonstationary random	Gaussian
Beat-like signals with period $\approx T$	Bartlett (triangular)
Narrow-band random	Rectangular
Stationary random	Hamming
Important low-level components mixed with widely spaced high-level spectral components	Parzen
Broad-band random (white noise, pink noise, etc.)	Gaussian

Characteristics of the signal that is being analyzed and also the nature of the system that generates the signal should be considered in choosing an appropriate truncation window. In particular, the Hanning window is recommended for signals generated by heavily damped systems and the Hamming window is recommended for use with lightly damped systems. Table 5.7 lists some useful signal types and appropriate window functions.

5.5.8 Spectral Approach to Process Monitoring

In mechanical systems, component degradation may be caused by vibrating excitations, which can result in malfunction or failure. In this sense, continuous monitoring during testing of mechanical deterioration in various critical components of a vibratory system is of prime importance. This usually cannot be done by simple visual observation, unless malfunction is detected by operability monitoring of the system. However, since mechanical degradation is always associated with a change in vibration level, by continuously monitoring the development of Fourier spectra in time (during system operation) at various critical locations of the system, it is possible to conveniently detect any mechanical deterioration and impending failure. In this respect, real-time Fourier analysis is very useful in process monitoring and failure detection and prediction. Special-purpose real-time analyzers with the capability of spectrum comparison (often done by an external command) are available for this purpose.

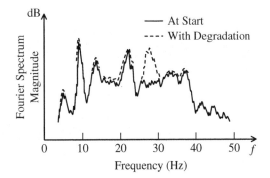

FIGURE 5.17 Effect of mechanical degradation on a monitored Fourier spectrum.

Various mechanical deteriorations manifest themselves at specific frequency values. A change in spectrum level at a particular frequency (and its multiples) would indicate a specific type of mechanical degradation or component failure. An example is given in Figure 5.17, which compares the Fourier spectrum at a monitoring location of a vibratory system at the start of test with the Fourier spectrum after some mechanical degradation has taken place. To facilitate spectrum comparison within a narrow-frequency band, it is customary to plot such Fourier spectra on a linear frequency axis. It is seen that the overall spectrum levels have increased as a result of mechanical degradation. Also, a significant change has occurred near 30 Hz. This information is useful in diagnosing the cause of degradation or malfunction. Figure 5.17 might indicate, for example, impending failure of a component having resonant frequency close to 30 Hz.

5.5.9   Cepstrum

A function known as the *cepstrum* is sometimes used to facilitate the analysis of Fourier spectrum in detecting mechanical degradation. The cepstrum (complex) $C(\tau)$ of a Fourier spectrum $Y(f)$ is defined by

$$C(\tau) = \mathfrak{F}^{-1} \log Y(f) \tag{5.54}$$

The independent variable τ is known as quefrency, and it has the units of time.

An immediate advantage of cepstrum arises from the fact that the logarithm of the Fourier spectrum is taken. From Equation 5.33 it is clear that, for a system having frequency-transfer function $H(f)$, and excited by a signal having Fourier spectrum $U(f)$, the response Fourier spectrum $Y(f)$ could be expressed in the logarithmic form:

$$\log Y(f) = \log H(f) + \log U(f) \tag{5.55}$$

Since the right-hand side terms are added rather than multiplied, any variation in $H(f)$ at a particular frequency will be less affected by a possible low-spectrum level in the excitation $U(f)$ at that frequency, when considering $\log Y(f)$ rather than $Y(f)$. Consequently, any degradation will be more conspicuous in the cepstrum than in the Fourier spectrum. Another advantage of cepstrum is that it is better capable of detecting variation in phenomena that manifest themselves as periodic components in the Fourier spectrum (for example, harmonics and sidebands). Such phenomena, which appear as repeated peaks in the Fourier spectrum, occur as a single peak in the cepstrum, and so any variations can be detected more easily.

5.6   Other Topics of Signal Analysis

In this section, we will briefly address some other important topics of signal analysis. We will start by discussing bandwidth in different contexts then we will present several practically useful analysis procedures and results on vibration signals.

5.6.1   Bandwidth

Bandwidth has different meanings depending on the particular context and application. For example, when studying the response of a dynamic system, the bandwidth relates to the fundamental resonant frequency, and correspondingly to the speed of response for a given excitation. In band-pass filters, the bandwidth refers to the frequency band within which the frequency components of the signal are allowed through the filter, the frequency components outside the band being rejected by it. With respect to measuring instruments, bandwidth refers to the range of frequencies within which the instrument measures a signal accurately. Note that these various interpretations of bandwidth are somewhat related. For example, if a signal passes through a band-pass filter, then we know that its frequency content is within the bandwidth of the filter; but we cannot determine the actual frequency content of the signal

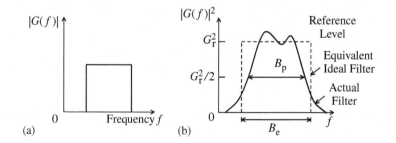

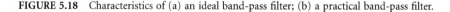

FIGURE 5.18   Characteristics of (a) an ideal band-pass filter; (b) a practical band-pass filter.

through such an observation. In this context, the bandwidth appears to represent a frequency uncertainty in the observation (i.e., the larger the bandwidth of the filter, the less certain the actual frequency content of a signal that is allowed through the filter).

5.6.2 Transmission Level of a Band-Pass Filter

Practical filters can be interpreted as dynamic systems. In fact all physical, dynamic systems (e.g., mechanical structures) are analog filters. It follows that the filter characteristics can be represented by the frequency-transfer function $G(f)$ of the filter. A magnitude squared plot of such a filter transfer function is shown in Figure 5.18. In a logarithmic plot the magnitude-squared curve is obtained by simply doubling the corresponding magnitude (Bode plot) curve. Note that the actual filter transfer function (Figure 5.18(b)) is not flat like the ideal filter shown in Figure 5.18(a). The reference level G_r is the average value of the transfer function magnitude in the neighborhood of its peak.

5.6.3 Effective Noise Bandwidth

Effective noise bandwidth of a filter is equal to the bandwidth of an ideal filter that has the same reference level and that transmits the same amount of power from a white noise source. Note that white noise has a constant (flat) PSD. Hence, for a noise source of unity PSD, the power transmitted by the practical filter is given by

$$\int_0^\infty |G(f)|^2 df$$

which, by definition, is equal to the power $G_r^2 B_e$ transmitted by the equivalent ideal filter. Hence, the effective noise bandwidth B_e is given by

$$B_e = \int_0^\infty |G(f)|^2 df / G_r^2 \tag{5.56}$$

5.6.4 Half-Power (or 3 dB) Bandwidth

Half of the power from a unity-PSD noise source, as transmitted by an ideal filter, is $G_r^2 B_e / 2$. Hence, $G_r/\sqrt{2}$ is referred to as the *half-power level*. This is also known as a 3 dB level because $20 \log_{10} \times \sqrt{2} = 10 \log_{10} 2 = 3$ dB. (*Note*: 3 dB refers to a power ratio of 2 or an amplitude ratio of $\sqrt{2}$. Furthermore, 20 dB corresponds to an amplitude ratio of 10 or a power ratio of 100). The 3 dB (or half-power) bandwidth corresponds to the width of the filter transfer function at the half-power level. This is denoted by B_p in Figure 5.18(b). Note that B_e and B_p are different in general. However, in an ideal case where the magnitude-squared filter characteristic has linear rise and fall-off segments, these two bandwidths are equal (see Figure 5.19).

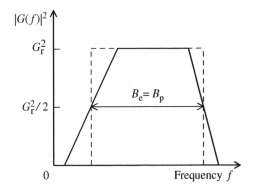

FIGURE 5.19 An idealized filter with linear segments.

5.6.5 Fourier Analysis Bandwidth

In Fourier analysis, bandwidth is interpreted, again, as the *frequency uncertainty* in the spectral results. In analytical FIT results, which assume that the entire signal is available for analysis, the spectrum is continuously defined over the entire frequency range $[-\infty, \infty]$ and the frequency increment df is infinitesimally small ($df \rightarrow 0$). There is no frequency uncertainty in this case, and the analysis bandwidth is infinitesimally narrow.

In digital Fourier transform, the discrete spectral lines are generated at frequency intervals of ΔF. This finite frequency increment ΔF, which is the frequency uncertainty, is therefore the analysis bandwidth B for this analysis. Note that $\Delta F = 1/T$, where T is the record length (or window length for a rectangular window). It follows also that the minimum frequency that has a meaningful accuracy is the bandwidth. This interpretation for analysis bandwidth is confirmed by noting the fact that harmonic components of frequency less than ΔF (or period greater than T) cannot be studied by observing a signal record of length less than T. Analysis bandwidth carries information regarding distinguishable minimum frequency separation in computed results. In this sense bandwidth is directly related to the frequency resolution of analyzed results. The accuracy of analysis increases by increasing the record length T (or decreasing the analysis bandwidth B).

When a time window other than the rectangular window is used to truncate a measured vibration signal, then reshaping of data occurs according to the shape of the window. This reshaping reduces leakage due to suppression of side lobes of the Fourier spectrum of the window. At the same time, however, an error is introduced due to the information lost through data reshaping. This error is proportional to the bandwidth of the window itself. The effective noise bandwidth of a rectangular window is only slightly less than $1/T$, because the main lobe of its Fourier spectrum is nearly rectangular. Hence, for all practical purposes, the effective noise bandwidth can be taken as the analysis bandwidth. Note that data truncation (multiplication in the time domain) is equivalent to convolution of the Fourier spectrum (in the frequency domain). The main lobe of the spectrum uniformly affects all spectral lines in the discrete spectrum of the data signal. It follows that a window main lobe having a broader bandwidth (effective noise bandwidth) introduces a larger error into the spectral results. Hence, in digital Fourier analysis, bandwidth is taken as the effective noise bandwidth of the time window that is employed.

5.6.6   Resolution in Digital Fourier Results

Resolution is the frequency separation between spectral lines in digital Fourier-analysis results. For a data record of length T, the resolution is $\Delta F = 1/T$ irrespective of the type of window used. There is a noteworthy distinction between analysis bandwidth and resolution. Suppose that we have a data record of length T. If we double the length by augmenting it with trailing zeros, digital Fourier analysis of the resulting record of length $2T$ will yield a spectral line separation of $1/(2T)$. Thus, the resolution is halved. But, unless the true signal value is also zero in the second time interval $t[T, 2T]$, no new information is present in the augmented record of duration $[0, 2T]$ in comparison to the original record of duration $[0, T]$. So, the analysis bandwidth (a measure of accuracy) will remain unchanged. If, on the other hand, the signal itself was sampled over $[0, 2T]$ and the resulting $2N$ data points were used in digital Fourier analysis, the bandwidth as well as the resolution would be halved.

Some relations that are useful in the digital computation of spectral results for signals are summarized in Box 5.3.

5.7   Overlapped Processing

Digital Fourier analysis is performed on blocks of sampled data (e.g., $2^{10} = 1024$ samples at a time). In overlapped processing, each data block is made to include part of the previous data block that was analyzed. After completing a computation, the overlapped data at the end of the computed block is moved to the beginning of the block, and the leading vacancy is filled with new data so that the end data in one block is identical to the beginning data in the next block, in the overlapped region. In other words, the overlapped portions of each data block (the two end portions) are processed twice. It follows that if there is 50% (or more) overlapping then the entire data block is processed twice. Three main reasons can

---

Box 5.3

USEFUL RELATIONS FOR DIGITAL SPECTRAL COMPUTATIONS

$$\mathfrak{F}\, y(t) \xrightarrow[\text{(FFT)}]{\text{DFT}} Y(f)$$ Fourier spectrum

$$\frac{1}{T} Y^*(f)Y(f)$$ = Power spectral density (PSD)

Power spectrum   = $B \times$ Power spectral density $= \dfrac{B}{T} Y^*(f)Y(f)$

Energy spectrum   = $T \times$ Power spectrum $= BY^*(f)Y(f)$

Energy spectral density $= \dfrac{1}{B} \times$ Energy spectrum $= Y^*(f)Y(f)$

RMS spectra $= \left[\dfrac{2}{B} \int_f^{f+B} |Y(f')|^2 \mathrm{d}f' \right]^{1/2}$ * one sided
(always shown for   * like $|Y(f)|$ but smoother
+ve frequencies only)   * no phase information
                                                                           * increase $B \rightarrow$ high
                                                                             bandwidth

Note:
$T = $ Record length
$B = $ Bandwidth of digital analysis (minimum frequency for which meaningful results are obtained) $\rightarrow$ includes
    window effect
Periodic or stationary signals   Use power spectra
(infinite energy)
Transient signals (finite energy)   Energy spectra can be used
One-sided spectrum $= 2 \times$ (+ve frequency part of two-sided spectrum)
Coherent output power $=$ coherence $\gamma_{uy}^2 \times$ output power $\leftarrow$ could be power spectrum (spectrum or spectral density)
or PSD of the output

---

be given for using overlapped processing in digital Fourier analysis:

1. It is an effective means of averaging spectral results.
2. It reduces the waiting time for assembling the data buffer.
3. It reduces the error caused by the end shaping effect of time windows (when a window other than the rectangle window is used).

From reasons 1 and 2, it is clear that, due to overlapping, the statistical error of computations is reduced for the same speed of computation, and the computing power is more efficiently used. To explain reason 3, let us examine Figure 5.20. This example shows a 50% overlap in data. It is seen that the window function can be assumed to be relatively flat, at least over 50% of the window length (record length). Then the entire data block will correspond to the flat part of the window in three successive analyses. Consequently, the shaping error (or the error due to increased analysis bandwidth) that is caused by a nonrectangular time window is virtually eliminated by overlapped processing. The flatness of a time window is determined by its effective noise bandwidth B_e. The effective record length T_e is defined as

$$T_e = \frac{1}{B_e} \qquad (5.57)$$

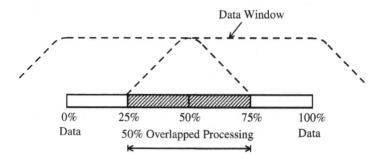

FIGURE 5.20 Overlapped processing of windowed signals.

which provides a measure for the flat segment of the window. The percentage effective record length is given by T_e as a percentage of the actual record length T. The degree of overlapping is chosen using the relation

$$\%\text{overlap} = 100\left(1 - \frac{T_e}{T}\right) \tag{5.58}$$

Example 5.3

For a Hamming window, $B_e = 1.4/T$. Hence, a typical value for the percentage overlap is

$$100\left(1 - \frac{1}{1.4}\right) = 29\%$$

We might want to use a conservative overlap and even go up to 50% in this case because the window is not quite flat.

5.7.1 Order Analysis

Speed related vibrations in rotation machinery may be analyzed through order analysis. Machinery vibrations under start-up (accelerating) and shut-down (decelerating) conditions are analyzed in this manner. Orders represent the rotating-speed-related frequency components in a response signal. The ratio of the response frequency to the rotating speed is termed "order."

Order analysis is done essentially through digital Fourier analysis of a rotating-speed-related response signal. Practically, this may be accomplished in many ways. The format in which the spectral results are presented will depend on the procedure used in order analysis. Some of the typical formats of data presentation are given below.

5.7.1.1 Speed-Spectral Map

As the rotating speed of a machine is changing in a given range, the Fourier spectrum of the response signal is determined for equal increments of speed. The results are presented as a speed spectral map which is a three-dimensional cascade diagram (or *waterfall display*). The two base axes of the plot are spectral frequency and rotating speed. The third axis gives the spectral magnitude (see Figure 5.21). These types of plots are useful in identifying order-related components during start-up or coast-down conditions. Note that for each speed the frequency band of digital Fourier analysis is kept the same (i.e., fixed sampling rate). Each distinct crest trace denotes an order-related resonance. The fact that these traces are almost straight lines indicates the significance of order (the ratio, frequency/rotating speed) in exciting these resonances.

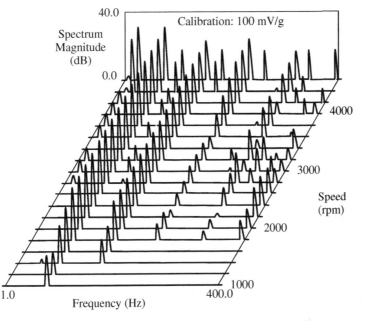

FIGURE 5.21 A speed-spectral map obtained from order analysis.

5.7.1.2   Time-Spectral Map

Under variable speed conditions (not necessarily accelerating or decelerating) the response signal is Fourier analyzed at equal increments of time. The results are plotted in a *cascade diagram*, with frequency and time as the base axes. The third axis again represents the magnitude of the Fourier spectrum (see Figure 5.22). In this case, the crest traces are not necessarily straight, and can change their orientation arbitrarily. This variation in crest orientation is determined by the degree of speed variation.

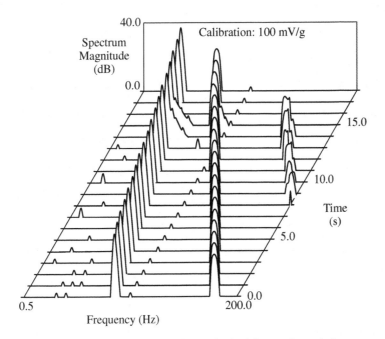

FIGURE 5.22 A time-spectral map obtained from order analysis.

5.7.1.3   Order Tracking

In order tracking, a "tracking frequency multiplier" monitors the rotating speed of the machine (as for a speed-spectral map). But, in the present case, the sampling rate of the response signal (for Fourier analysis) is changed in proportion to the rotating speed. Note that, in this manner, the maximum useful frequency (approximately 400/512 × Nyquist Frequency) is increased as the rotating speed increases, so that the aliasing effects are reduced. If the same sampling rate is used for high speeds (as in the Speed-Spectral Map discussed above), aliasing error can be significant at high rotating speeds.

In presenting order tracking spectral results, the frequency axis is typically calibrated in orders. Both speed-spectral maps and time-spectral maps may be presented in this manner. Other types of data presentation may be used as well in order analysis. For example, instead of the Fourier spectrum of the response signal, power spectrum or composite power spectrum (in which the total signal power is computed in specified frequency bands and presented as a function of the rotating speed) may be used in the schemes described in this section.

Order analysis provides information on most severe operating speeds with respect to vibration (and dynamic stress). For example, suppose that, for a given speed of operation, two major resonances occur, one at 10 Hz and the other at 80 Hz. Then, the structure of the system (rotating machine and its support fixtures) should be modified to change and preferably damp out these resonances. Furthermore, the most desirable operating speed can be chosen in terms of the lowest resonant peaks by observing a speed-spectral map.

Bibliography

Bendat, J.S. and Piersol, A.G. 1971. *Random Data: Analysis and Measurement Procedures*, Wiley-Interscience, New York.

Brigham, E.O. 1974. *The Fast Fourier Transform*, Prentice Hall, Englewood Cliffs, NJ.

Broch, J.T. 1980. *Mechanical Vibration and Shock Measurements*, Bruel and Kjaer, Naerum.

de Silva, C.W. 1983. *Dynamic Testing and Seismic Qualification Practice*, D.C. Heath and Co., Lexington, MA.

de Silva, C.W., Optimal estimation of the response of internally damped beams to random loads in the presence of measurement noise, *J. Sound Vib.*, 47, 4, 485–493, 1976.

de Silva, C.W., The digital processing of acceleration measurements for modal analysis, *Shock Vib. Dig.*, 18, 10, 3–10, 1986.

de Silva, C.W. 2006. *Vibration — Fundamentals and Practice*, 2nd Edition, Taylor & Francis, CRC Press, Boca Raton, FL.

de Silva, C.W. 2005. *Mechatronics — An Integrated Approach*, Taylor & Francis, CRC Press, Boca Raton, FL.

Ewins, D.J. 1984. *Modal Testing: Theory and Practice*, Research Studies Press Ltd, Letchworth.

MATLAB *Control Systems Toolbox*, The Math Works, Inc., Natick, MA, 2004.

Meirovitch, L. 1980. *Computational Methods in Structural Dynamics*, Sijthoff & Noordhoff, Rockville, MD.

Randall, R.B. 1977. *Application of B&K Equipment to Frequency Analysis*, Bruel and Kjaer, Naerum.

6

Wavelets — Concepts and Applications

Pol D. Spanos
Rice University

Giuseppe Failla
*Università degli Studi
Mediterranea di Reggio Calabria*

Nikolaos P. Politis
Rice University

Summary

Section 6.1 provides a brief introduction to wavelet concepts in vibration-related applications. Aspects of time–frequency analysis are discussed in Section 6.2. Specifically, the Gabor and wavelet transforms are outlined. Further, several wavelet families commonly used in vibration-related applications are presented. Estimation of time-dependent spectra of stochastic processes is considered in Section 6.3. Section 6.4 to Section 6.7 discuss applications of wavelet analysis in vibration-related applications. In particular, applications in random field simulation, system identification, damage detection, and material characterization are examined. Section 6.8 provides an overview and concluding remarks on the applicability and usefulness of the wavelet analysis in vibration theory. To enhance the usefulness of this chapter, a list of readily available references in the form of books and archival articles is provided.

6.1 Introduction

Wavelets-based representations offer an important option for capturing localized effects in many signals. This is achieved by employing representations *via* double integrals (continuous transforms), or *via* double series (discrete transforms). Seminal to these representations are the processes of *scaling* and *shifting* of a generating (*mother*) function. Over a period of several decades, wavelet analysis has been set on a rigorous mathematical framework and has been applied to quite diverse fields. Wavelet families associated with specific mother functions have proven quite appropriate for a variety of problems. In this context, fast decomposition and reconstruction algorithms ensure computational efficiency, and rival classical spectral analysis algorithms such as the fast Fourier transform (FFT). The field of vibration analysis has benefited from this remarkable mathematical development in conjunction with vibration monitoring, system identification, damage detection, and several other tasks. There is a voluminous body of literature focusing on wavelet analysis. However, this chapter has the restricted objective of, on one hand, discussing concepts closely related to vibration analysis, and on the other hand, citing sources that can be readily available to a potential reader. In view of this latter objective, almost exclusively books and archival articles are included in the list of references. First, theoretical concepts are briefly presented; for more mathematical details, the reader may consult references [1–23]. Next, the theoretical concepts are supplemented by vibration-analysis-related sections on time-varying spectra estimation, random field synthesis, structural identification, damage detection, and material characterization. It is noted that most of the mathematical developments pertain to the interval [0,1] relating to dimensionless independent variables derived by normalization with respect to the spatial or temporal "lengths" of the entire signals.

6.2   Time–Frequency Analysis

A convenient way to introduce the wavelet transform is through the concept of time–frequency representation of signals. In the classical Fourier theory, a signal can be represented either in the time or in the frequency domain, and the Fourier coefficients define the average spectral content over the entire duration of the signal. The Fourier representation is appropriate for signals that are stationary, in terms of parameters which are deemed important for the problem in hand, but becomes inadequate for nonstationary signals, in which important parameters may evolve rapidly in time.

The need for a time–frequency representation is obvious in a broad range of physical problems, such as acoustics, image processing, earthquake and wind engineering, and a plethora of others. Among the time–frequency representations available to date, the wavelet transform has unique features in terms of efficacy and versatility. In mathematical terms, it involves the concept of *scale* as a counterpart to the concept of frequency in the Fourier theory. Thus, it is also referred to as time-scale representation. Its formulation stems from a generalization of a previous time–frequency representation, known as the Gabor transform. For completeness, and to underscore the significant advantages achieved by the development of the wavelet transform, the Gabor transform is briefly discussed in Section 6.2.1. Section 6.2.2 is entirely devoted to the wavelet transform, and the most commonly used wavelet families are described in Section 6.2.3.

6.2.1   Gabor Transform

The first steps in time–frequency analysis trace back to the work of Gabor [24], who applied in signal analysis fundamental concepts developed in quantum mechanics by Wigner a decade earlier [25]. Given a function $f(t)$ belonging to the space of finite-energy one-dimensional functions, denoted by $L^2(\mathbb{R})$, Gabor introduced the transform

$$G_f(\omega, t_0) = \int_{-\infty}^{\infty} f(t)\overline{g(t - t_0)}\, e^{-i\omega(t - t_0)}\, dt \tag{6.1}$$

where $g(t)$ is a window and the bar (¯) denotes complex conjugation. This transform, generally referred to as the *continuous Gabor transform* (CGT) or the short-time Fourier transform of $f(t)$, is a complete representation of $f(t)$. That is, the original function $f(t)$ can be reconstructed as

$$f(t) = \frac{1}{2\pi\|g\|^2} \int_{-\infty}^{\infty} \int_{-\infty}^{\infty} G_f(\omega, t_0) g(t - t_0) e^{i\omega(t - t_0)}\, d\omega\, dt_0 \tag{6.2}$$

where $\|g\|^2 = \int_{-\infty}^{\infty} |g(t)|^2\, dt$. The Gabor transform (Equation 6.1) may be seen as the projection of $f(t)$ onto the family $\{g_{(\omega, t_0)}(t); \omega, t_0 \in \mathbb{R}\}$ of *shifted* and *modulated* copies (atoms) of $g(t)$ expressed in the form

$$g_{(\omega, t_0)}(t) = e^{i\omega(t - t_0)} g(t - t_0) \tag{6.3}$$

These time–frequency atoms, also referred to as Gabor functions, are shown in Figure 6.1 for three different values of ω. Clearly, if $g(t)$ is an appropriate window function, then Equation 6.1 may be regarded as the standard Fourier transform of the function $f(t)$, localized at the time t_0. In this context, t_0 is the

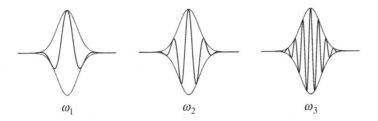

ω_1 ω_2 ω_3

FIGURE 6.1   Plots of Gabor function $g(\omega, t_0)$ versus the independent variable x for three values of the frequency ω; the effective support is the same for the three values of the frequency.

time parameter which gives the center of the window, and ω is the frequency parameter which is used to compute the Fourier transform of the windowed signal.

As intuition suggests, the accuracy of the CGT representation (Equation 6.2) of $f(t)$ depends on the window function $g(t)$, which must exhibit good localization properties both in the time and the frequency domains. As discussed in Ref. [6], a measure of the localization properties may be obtained by the average and the standard deviation of the density $|g(t)|^2$ in the time domain. That is,

$$\langle t \rangle = \int_{-\infty}^{\infty} t|g(t)|^2 \, dt \tag{6.4a}$$

$$\sigma_t^2 = \int_{-\infty}^{\infty} (t - \langle t \rangle)^2 |g(t)|^2 \, dt \tag{6.4b}$$

The counterparts of Equation 6.4a and Equation 6.4b in the frequency domain are

$$\langle \omega \rangle = \int_{-\infty}^{\infty} \omega|\hat{G}(\omega)|^2 \, d\omega \tag{6.5a}$$

$$\sigma_\omega^2 = \int_{-\infty}^{\infty} (\omega - \langle \omega \rangle)^2 |\hat{G}(\omega)|^2 \, d\omega \tag{6.5b}$$

where $\hat{G}(\omega)$ denotes the Fourier transform of $g(t)$ given by the equation

$$\hat{G}(\omega) = \frac{1}{\sqrt{2\pi}} \int_{-\infty}^{\infty} g(t)e^{-i\omega t} \, dt \tag{6.6}$$

The well-known *Heisenberg uncertainty principle* is in actuality a mathematically proven property and states that the values σ_t and σ_ω cannot be independently small [6]. Specifically, for an arbitrary window normalized so that $\|g\|^2 = 1$, it can be shown that

$$\sigma_t \sigma_\omega \geq \frac{1}{2} \tag{6.7}$$

Thus, high resolution in the time domain (small value of σ_t) may be generally achieved only at the expense of a poor resolution (bigger than a minimum value σ_ω) in the frequency domain and *vice versa*. Note that the optimal time–frequency resolution, that is $\sigma_t \sigma_\omega = 1/2$, may be attained when the Gaussian window

$$g(t) = \frac{1}{\sqrt[4]{2\pi\sigma_t^2}} \exp\left(-\frac{t^2}{4\sigma_t^2}\right) \tag{6.8}$$

is selected.

Clearly, as a time–frequency representation, the Gabor transform exhibits considerable limitations. The time support, governed by the window function $g(t)$, is equal for all of the Gabor functions (Equation 6.3) for all frequencies (see Figure 6.1). In order to achieve good localization of high-frequency components, narrow windows are required; as a result of that, low-frequency components are poorly represented. Thus, a more flexible representation with nonconstant windowing is quite desirable, to enhance the time resolution for short-lived high-frequency phenomena and frequency resolution for long-lasting low-frequency phenomena.

6.2.2 Wavelet Transform

The preceding shortcomings of the Gabor transform have been overcome with significant effectiveness and efficiency by wavelets-based signal representation. Its two formulations, continuous and discrete, are described in the ensuing sections. Because of the numerous applications of wavelets beyond time–frequency analysis, the *t*-time domain will be replaced by a generic *x*-space domain. For succinctness, the formulation will be developed for scalar functions only, but generalization for multidimensional spaces is well established in the literature [1–22].

6.2.2.1 Continuous Wavelet Transform

The concept of wavelet transform was introduced first by Goupillaud et al. for seismic records analysis [26, 27]. In analogy to the Gabor transform, the idea consists of decomposing a function $f(x)$ into a two-parameter family of elementary functions, each derived from a basic or *mother wavelet*, $\psi(x)$. The first parameter, *a*, corresponds to a dilation or compression of the mother wavelet that is generally referred to as *scale*. The second parameter, *b*, determines a *shift* of the mother wavelet along the *x*-domain. In mathematical terms

$$W_f(a,b) = \frac{1}{\sqrt{a}} \int_{-\infty}^{\infty} f(x)\overline{\psi\left(\frac{x-b}{a}\right)}\,dx \qquad (6.9)$$

where $a \in \mathbb{R}^+$, $b \in \mathbb{R}$. In the literature, Equation 6.9 is generally referred to as *continuous wavelet transform* (CWT). Note that the factor $a^{-1/2}$ is a normalization factor, included to insure that the mother wavelet and any dilated wavelet $a^{-1/2}\psi(x/a)$ have the same total energy [26]. Clearly, alternative normalizations may also be chosen [1].

An example of wavelet functions is shown in Figure 6.2 for different values of the scale parameter *a*. As a result of scaling, all the wavelet functions exhibit the same number of cycles within the *x*-support of the mother wavelet. Obviously, the spatial and frequency localization properties of the wavelet transform depend on the value of the parameter *a*. As *a* approaches zero, the dilated wavelet $a^{-1/2}\psi(x/a)$ is highly concentrated at the point $x = 0$; the wavelet transform, $W_f(a,b)$, then gives increasingly sharper spatial resolution displaying the small-scale/higher-frequency features of the function $f(x)$, at various locations *b*. However, as *a* approaches $+\infty$, the wavelet transform $W_f(a,b)$ gives increasingly coarser spatial resolution, displaying the large-scale/low-frequency features of the function $f(x)$.

For the function $f(x)$ to be reconstructable from the set of coefficients (Equation 6.9), in the form

$$f(x) = \frac{1}{\pi c_\psi} \int_0^\infty \int_{-\infty}^\infty W_f(a,b)\psi_{a,b}(x)\frac{da}{a^2}\,db \qquad (6.10)$$

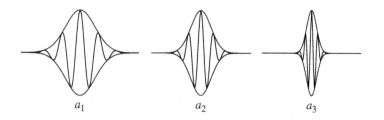

$$a_1 \qquad\qquad a_2 \qquad\qquad a_3$$

FIGURE 6.2 Plots versus time of wavelet functions corresponding to three different values of a scale *a* of the same mother function; the effective time support increases with the magnitude of the scale.

where $\psi_{a,b}(x) = a^{-1/2}\psi[(x-b)/a]$, the wavelet function $\psi(\cdot)$ must satisfy the admissibility condition

$$c_\psi = \int_{-\infty}^{\infty} \frac{|\hat{\Psi}(\omega)|^2}{|\omega|}\, d\omega < \infty \tag{6.11}$$

where $\hat{\Psi}(\omega)$ denotes the Fourier transform of $\psi(x)$. As pointed out in Ref. [26], the condition 6.11 includes a set of subconditions, such as:

1. The analyzing wavelet $\psi(\cdot)$ is absolutely integrable and square integrable. That is,

$$\int_{-\infty}^{\infty} |\psi(x)|\, dx < \infty \tag{6.12a}$$

$$\int_{-\infty}^{\infty} |\psi(x)|^2\, dx < \infty \tag{6.12b}$$

2. The Fourier transform $\hat{\Psi}(\omega)$ must be sufficiently small at the vicinity of the origin $\omega = 0$, or in mathematical terms

$$\int_{-\infty}^{\infty} \frac{|\hat{\Psi}(\omega)|}{|\omega|}\, d\omega < \infty \tag{6.13}$$

Subcondition 2, then, implies that $\hat{\Psi}(0) = 0$; that is, $\int_{-\infty}^{\infty} \psi(x)dx = 0$. Therefore, for an analyzing wavelet to be admissible, its real and imaginary parts must both be symmetric with respect to the x-axis. From the reconstruction formula (Equation 6.10), it can be shown that

$$\|f\|^2 = \frac{1}{\pi c_\psi} \int_0^\infty \int_{-\infty}^\infty |W_f(a,b)|^2 \frac{da}{a^2}\, db \tag{6.14}$$

Based on Equation 6.14, the square modulus of the wavelet transform (Equation 6.9) is often taken as an energy density in a spatial-scale domain. Extensive use of this concept has been made for spectra estimation purposes, as discussed in Section 6.3.

Note that the reconstruction wavelet in Equation 6.10 can be different from the analyzing wavelet used in Equation 6.9. That is, under some admissibility conditions on $\chi(x)$ [1], the original function $f(x)$ may be reconstructed as

$$f(x) = \frac{1}{c_{\psi\chi}} \int_0^\infty \int_{-\infty}^\infty W_f(a,b)\chi_{a,b}(x) \frac{da}{a^2}\, db \tag{6.15}$$

where $\chi_{a,b}(x) = a^{-1/2}\chi[(x-b)/a]$ and $c_{\psi\chi}$ is a constant parameter depending on the Fourier transforms of both $\psi(x)$ and $\chi(x)$. This property, referred to as redundancy in mathematical terms, may be advantageous in some applications for reducing the error due to noise in signal reconstruction [28, 29], but highly undesirable for signal coding or compression purposes [1]. Further, under certain conditions [1], the following simplified reconstruction formula holds

$$f(x) = \frac{1}{k_\psi} \int_0^\infty W_f(a,x) \frac{da}{a^{3/2}} \tag{6.16}$$

where k_ψ is a constant parameter given by the equation

$$k_\psi = \sqrt{2\pi} \int_0^\infty \frac{\overline{\hat{\Psi}(\omega)}}{\omega}\, d\omega \tag{6.17}$$

Use of this formula has been made, in a discrete version, in the approximation theory of functional spaces [1] and also in structural identification applications, as discussed in Section 6.5.

6.2.2.2 Discrete Wavelet Transform

For numerical applications, where fast decomposition or reconstruction algorithms are generally required, a discrete version of the CWT is to prefer. In this sense, a natural way to define a *discrete wavelet transform* (DWT) is

$$W_f(j,k) = \frac{1}{\sqrt{a_0^j}} \int_{-\infty}^{\infty} f(x)\overline{\psi(a_0^{-j}x - kb_0)}dx, \qquad j,k \in \mathbb{Z} \tag{6.18}$$

Equation 6.18 is derived from a straightforward discretization of the CWT (Equation 6.9) by considering the discrete lattice $a = a_0^j$, $a_0 > 1$, $b = kb_0a_0^j$, $b_0 \neq 0$. In developing Equation 6.18, however, the main mathematical concern is to ensure that *sampling* the CWT on a discrete set of points does not lead to a loss of information about the wavelet-transformed function $f(x)$. Specifically, the original function $f(x)$ must be fully recoverable from a discrete set of wavelet coefficients. That is,

$$f(x) = \sum_{j,k \in \mathbb{Z}} W_f(j,k)\psi_{j,k}(x) \tag{6.19}$$

where $\psi_{j,k}(x) = a_0^{-j/2}\psi(a_0^{-j}x - kb_0)$. Another crucial aspect in Equation 6.18 involves selecting the wavelet functions $\psi_{j,k}(x)$ such that Equation 6.19 may be regarded as the expansion of $f(x)$ in a *basis*, thus eliminating the redundancy of the CWT.

This issues are addressed by using the theory of Hilbert space *frames*, introduced in 1952 by Duffin and Schaeffer in context with non-harmonic Fourier series [30]. In general, if $h_\lambda(x) \in L^2(\mathbb{R})$ and Λ is a countable set, a family of functions $\{h_\lambda(x); \lambda \in \Lambda\}$ constitutes a frame, if for any $f(x) \in L^2(\mathbb{R})$

$$A\|f\|^2 \leq \sum_{\lambda \in \Lambda} |\langle f, h_\lambda \rangle|^2 \leq B\|f\|^2 \tag{6.20}$$

where $\langle f, h_\lambda \rangle = \int_{-\infty}^{\infty} f(x)\overline{h_\lambda(x)}\,dx$ and $A > 0, B < \infty$, the so-called *frame bounds*, are independent of $f(x)$ [1]. The concept of frame may be interpreted as an extension of the concept of basis, in the sense that the reconstruction of the original function is possible *via* stable numerical expressions in terms of the set $\{h_\lambda(x); \lambda \in \Lambda\}$. For instance, if the frame is *tight*, that is $A = B$, the simple formula

$$f(x) = \sum_{\lambda \in \Lambda} \langle f, h_\lambda \rangle h_\lambda(x) \tag{6.21}$$

holds [29].

In contrast to a basis, however, the vectors of a frame may be linearly dependent and, for this, a certain degree of redundancy is still retained in the reconstruction formula (Equation 6.21) [29, 31].

The concept of frame has played a crucial role in the formulation of the DWT. The first wavelet frames were constructed by Daubechies et al. [32]. Later, Battle [33] constructed orthonormal bases with an exponential decay. The ensemble of these results has demonstrated the advantages of the wavelet transform over the Gabor transform. In fact, it has been shown that discrete versions of Gabor transform are not capable of generating orthonormal bases [32] due to the so-called Balian–Low phenomenon [1].

Mallat [34] has shown that the orthonormal wavelet bases proposed by Battle can all be derived by a *multiresolution analysis*. The latter involves representing an arbitrary $f(x) \in L^2(\mathbb{R})$ as a limit of successive approximations, at different resolutions. That is, if $\{V_j\}_{j \in \mathbb{Z}}$ is a sequence of subspaces of $L^2(\mathbb{R})$, and f_j is the orthogonal projection of $f(x)$ on V_j, in a multiresolution analysis the following conditions hold

$$\lim_{j \to -\infty} f_j = f \tag{6.22a}$$

$$\lim_{j \to \infty} f_j = 0 \tag{6.22b}$$

Each approximation f_j, then, represents a smoothed version of $f(x)$ and, in the limit, more and more localized smoothing functions lead to the function $f(x)$. From a mathematical point of view [29, 31], a multiresolution analysis requires that

1. The subspaces V_j's are closed and embedded, that is

$$\cdots \subset V_2 \subset V_1 \subset V_0 \subset V_{-1} \subset V_{-2} \subset \cdots \tag{6.23}$$

where $V_{-m} \to L^2(\mathbb{R})$ for $m \to \infty$ and $f \in V_m \Leftrightarrow f(2\cdot) \in V_{m-1}$.

2. A scaling function $\phi(x) \in L^2(\mathbb{R})$ exists, such that, for each j, the family of functions

$$\phi_{j,k}(x) = 2^{-j/2}\phi(2^{-j}x - k), \quad k \in \mathbb{Z} \tag{6.24}$$

spans the subspace V_j and constitutes a Riesz basis for V_j, that is, there exists $0 < C' \leq C'' < \infty$ such that

$$C' \sum_k |c_k|^2 \leq \int_{-\infty}^{\infty} \left| \sum_k c_k \phi_{j,k}(x) \right|^2 dx \leq C'' \sum_k |c_k|^2 \tag{6.25}$$

for all sequences of numbers $(c_k)_{k \in \mathbb{Z}}$. Equation 6.25 is a more stringent condition of Equation 6.20 and includes the latter as a special case.

The concept of multiresolution analysis offers a straightforward and mathematically coherent approach to discrete wavelet analysis. Given a scaling function $\phi(x)$ as in 2, in fact, families of orthonormal wavelet bases

$$\psi_{j,k}(x) = 2^{-j/2}\psi(2^{-j}x - k), \quad j, k \in \mathbb{Z} \tag{6.26}$$

can be developed by appropriate algorithms. For this, Mallat has used the frequency response of a high-pass filter [35], while Daubechies has devised a systematic approach to build orthonormal wavelet bases with compact support in the x-domain [36]. Specifically, for each even integer $2M$, the Daubechies scaling function $\phi(x)$ can be computed as

$$\phi(x) = \sqrt{2} \sum_{k=0}^{2M-1} h_{k+1}\phi(2x - k) \tag{6.27}$$

where h_k's are $2M$ coefficients obtained by imposing M orthogonality conditions and M accuracy conditions to enhance the rate of convergence of the approximation to the original function $f(x)$. In turn, the mother wavelet is related to the scaling function $\phi(x)$ by the equation

$$\psi(x) = \sqrt{2} \sum_{k=0}^{2M-1} g_{k+1}\phi(2x - k) \tag{6.28}$$

where g_k's are the same as h_k's but reversed in order and with alternate signs. Numerical values of both series h_k's and g_k's are readily available in the literature [16].

Also based on multiresolution analysis concepts, a wavelet decomposition algorithm for image analysis has been developed [34, 35]. If associated to Daubechies wavelets, the algorithm becomes quite efficient from a computational point of view, since no numerical integration is involved to compute wavelet and scale coefficients. It relies on the projection of $f(x)$ onto a sufficiently fine scale j of the set 6.24. That is,

$$f(x) \approx f_j(x) = \sum_k c_k^j \phi_{j,k}(x) \tag{6.29}$$

where, for orthogonal wavelets,

$$c_k^j = \int_{-\infty}^{\infty} f(x)\phi_{j,k}(x)dx \tag{6.30}$$

Based on Equation 6.22a and Equation 6.22b, the projection $f_j(x)$ can be rewritten in terms of the projection $f_{j+1}(x)$ onto the coarser scale $(j + 1)$ and the incremental *detail* $\delta_{j+1}(x)$, that is the pieces of information contained in the subspace V_j and lost when "moving" to the subspace V_{j+1}. Therefore,

$$f_j(x) = f_{j+1}(x) + \delta_{j+1}(x) = f_{j+l}(x) + \delta_{j+1}(x) + \cdots + \delta_{j+l}(x) \approx \delta_{j+1}(x) + \cdots + \delta_{j+l}(x) \tag{6.31}$$

As a fundamental result of multiresolution analysis, the details $\delta_j(x)$ can be decomposed in terms of the set of wavelet functions at the same scale. That is,

$$\delta_j(x) = \sum_k d_k^j \psi_{j,k}(x) \tag{6.32}$$

where d_k^j's are the wavelet coefficients of $f(x)$. Based on Equation 6.28, both wavelet and scale coefficients can be computed recursively by the closed-form expressions

$$c_k^j = \sum_{l=0}^{2M-1} h_{l+1} c_{2k+l-1}^{j-1} \tag{6.33a}$$

$$d_k^j = \sum_{l=0}^{2M-1} g_{l+1} c_{2k+l-1}^{j-1} \tag{6.33b}$$

Similarly, the reconstruction algorithm can be implemented by the formula

$$c_k^{j-1} = \sum_l h_{k-2l+2} c_l^j + g_{k-2l+2} d_l^j \tag{6.34}$$

The reconstruction algorithm described by Equation 6.34 lends itself to interpretation as a scale linear system [37, 38]. Based on this concept, applications have also been developed for random field simulation [39].

6.2.3 Wavelet Families

A great number of wavelet families with various properties are available. Selecting an optimal family for a specific problem is not, in general, an easy task and there are properties that prove more important to certain fields of application. For instance, symmetry may be of great help for preventing dephasing in image processing, while regularity is critical for building smooth reconstructed signals or accurate nonlinear regression estimates. Compactly supported wavelets, either in the time or in the frequency domain, may be preferable for enhanced time or frequency resolution. The number of vanishing moments, M, that is the highest integer m for which the equation

$$\int_{-\infty}^{\infty} x^m \psi(x)dx = 0, \quad m = 0, 1, \ldots, M - 1 \tag{6.35}$$

holds is important in signal processing for compression, or in damage detection for enhancement of singularities in the vibration modes. Also, in some cases, wavelets may be required to be *progressive*. In mathematical terms, this means that their Fourier transform is defined only for positive frequencies. That is,

$$\hat{\Psi}(\omega) = 0, \quad \text{for } \omega < 0 \tag{6.36}$$

The progressive wavelet transform of a real-valued signal $f(t)$ and the associated *analytic signal*

$$z_f(t) = f(t) + iH[f(t)] \tag{6.37}$$

are related by the equation

$$W_f(a, b) = \frac{1}{2} W_{z_f}(a, b) \tag{6.38}$$

where $H[\cdot]$ denotes the Hilbert transform operator [40]. Equation 6.38 is quite useful for structural identification. Note also that significant reduction of computational costs is generally achieved if orthogonal wavelets in the frequency or in the x-domain are used.

A brief description of the most-used families is given below. A distinction is made between real and complex wavelets, and the most relevant properties for application purposes are discussed. A more exhaustive review may be in found in Ref. [15].

6.2.3.1 Real Wavelets

1. *Daubechies orthonormal wavelets* — A family of bases, each corresponding to a particular value of the parameter M in Equation 6.27 and Equation 6.28 [36]. Closed-form expressions for $\phi(x)$ in Equation 6.27 are available only for $M = 1$, to which the well-known Haar basis corresponds. In this case, the scaling function and the mother wavelet are

$$
\phi(x) = \begin{cases} 1, & 0 \le x < 1, \\ 0, & \text{elsewhere,} \end{cases}
\qquad
\psi(x) = \begin{cases} 1, & 0 \le x < \dfrac{1}{2}, \\ -1, & \dfrac{1}{2} \le x < 1, \\ 0, & \text{elsewhere.} \end{cases}
\tag{6.39}
$$

Various algorithms, however, are available in the literature for determining $\phi(x)$ and $\psi(x)$ numerically for $M > 1$.

Daubechies wavelets support both CWT and DWT, although the latter is most generally performed due to the fast decomposition and reconstruction algorithm mentioned in Section 6.2.2.2. Both $\phi(x)$ and $\psi(x)$ are compactly supported in the x-domain, and the support is equal to the segment $[0; 2M - 1]$. Also, M is equal to the number of vanishing moments of the wavelet function. Note that most Daubechies wavelets are not symmetric; regularity and harmonic-like shape increases with M.

2. *Meyer wavelets* — Families of wavelets [10], each defined for a particular choice of an auxiliary function $v(\omega)$ which appears in the following expression for the Fourier transform of the mother wavelet:

$$
\hat{\Psi}(\omega) = \begin{cases} \dfrac{1}{\sqrt{2\pi}} e^{i\omega/2} \sin\left[\dfrac{\pi}{2} v\left(\dfrac{3}{2\pi}|\omega| - 1 \right) \right], & \dfrac{2}{3}\pi \le |\omega| \le \dfrac{4}{3}\pi, \\[2ex] \dfrac{1}{\sqrt{2\pi}} e^{i\omega/2} \cos\left[\dfrac{\pi}{2} v\left(\dfrac{3}{4\pi}|\omega| - 1 \right) \right], & \dfrac{4}{3}\pi \le |\omega| \le \dfrac{8}{3}\pi, \\[2ex] 0, & |\omega| \notin \left[\dfrac{2}{3}\pi, \dfrac{8}{3}\pi \right], \end{cases}
\tag{6.40}
$$

for $v(\omega)$ to be an admissible auxiliary function it is required that

$$
v(\omega) = \begin{cases} 0, & \omega \le 0, \\ 1, & \omega \ge 1, \end{cases}
\tag{6.41a}
$$

$$
v(\omega) + v(1 - \omega) = 1, \quad 0 \le \omega \le 1
\tag{6.41b}
$$

The most common form of $v(\omega)$ in the literature is

$$
v(\omega) = \omega^4(35 - 84\omega + 70\omega^2 - 20\omega^3), \quad 0 \le \omega \le 1
\tag{6.42}
$$

The mother wavelet, for which only numerical expressions are available, is then constructed by inverse Fourier-transforming Equation 6.40.

Meyer wavelets are suitable for both CWT and DWT. Unlike Daubechies wavelets, they are compact in the frequency domain but not in the x-domain. Because of their fast decay, however, an effective x-support $[-8,8]$ is generally assumed. Appealing features of Meyer wavelets are orthogonality, infinite regularity, and symmetry.

3. *Mexican Hat wavelets* — A family of wavelets in the x-domain [15] related to a mother function that is proportional to the second derivative of the Gaussian probability density function.

That is,

$$\psi(x) = \frac{2}{\sqrt{3}} \pi^{-1/4}(1 - x^2)e^{-x^2/2} \tag{6.43}$$

The Mexican Hat wavelets allow CWT only. Unlike Daubechies or Meyer wavelets, Mexican Hat wavelets are not compact both in the frequency and in the x-domain, although an effective support $[-5,5]$ may be considered for practical calculations. They are infinitely regular and symmetric.

4. *Biorthogonal wavelets* — Families of wavelets derived by generalizing the ordinary concept of wavelet bases, and creating a pair of *dual wavelets*, say $(\psi(x), \tilde{\psi}(x))$, satisfying the following properties [41, 42]:

$$\int_{-\infty}^{\infty} \psi_{j,k}(x)\tilde{\psi}_{j',k'}(x)dx = \delta_{jj'}\delta_{kk'} \tag{6.44}$$

where the symbol δ_{mn} denotes the Kronecker delta. One wavelet, say $\psi(x)$, may be used for reconstruction and the dual one, $\tilde{\psi}(x)$, for decomposition. Therefore, Equation 6.18 and Equation 6.19 can be rewritten as

$$W_f(j, k) = 2^{-j/2} \int_{-\infty}^{\infty} f(x)\psi_{j,k}(2^{-j}x - k)dx, \quad j, k \in \mathbb{Z} \tag{6.45}$$

$$f(x) = \sum_{j,k \in \mathbb{Z}} W_f(j, k)\tilde{\psi}_{j,k}(x) \tag{6.46}$$

Biorthogonal wavelets support both CWT and DWT. The properties of a biorthogonal basis are specified in terms of a pair of integers (N_d, N_r). These integers, in analogy with the Daubechies wavelets, govern the regularity and the number of vanishing moments N_d of the decomposition wavelet $\psi(x)$, and the regularity and the number of vanishing moments N_r of the reconstruction wavelet $\tilde{\psi}(x)$. Obviously, this feature allows a greater number of choices for signal decomposition and reconstruction. Both wavelet functions $\psi(x)$ and $\tilde{\psi}(x)$ are symmetric.

6.2.3.2 Complex Wavelets

5. *Harmonic wavelets* — A Family of bases defined in the frequency domain by the formula [16, 43, 44]:

$$\hat{\Psi}_{m,n}(\omega) = \begin{cases} \dfrac{1}{2\pi(n - m)}, & m\pi \le \omega \le n\pi, \\ 0, & \text{elsewhere,} \end{cases} \tag{6.47}$$

where m and n are positive numbers but not necessarily integers. The pair of values m, n is referred to as *level m, n* and represents, for harmonic wavelets, the scale index j. A harmonic wavelet basis thus corresponds to a complete set of adjacent levels m, n, spanning all the positive frequency axis. By inverse-Fourier transforming Equation 6.47, the corresponding wavelet functions at a generic step k on the x-domain take the complex form:

$$\psi_{m,n,k}(x) = \frac{\exp\left[in2\pi\left(x - \dfrac{k}{n - m}\right)\right] - \exp\left[im2\pi\left(x - \dfrac{k}{n - m}\right)\right]}{i2\pi(n - m)\left(x - \dfrac{k}{n - m}\right)} \tag{6.48}$$

A common choice for the pairs m, n is $m, n = 0, 1; 2, 4; \ldots; 2^j, 2^{j+1}; \ldots$. In this case, all the wavelets have octave bands, except for the first one.

Harmonic wavelets have been devised in context with a DWT, for which extremely fast decomposition and reconstruction algorithms exist. They exhibit a compact support in the frequency domain (see Equation 6.47), while in the x-domain their rate of decay away from the wavelet's center is relatively low and proportional to x^{-1}. Further, they satisfy relevant orthogonality properties [16].

From Equation 6.48, it is seen that the real part of the wavelet is even, while the imaginary part is odd. For signal processing, this ensures that harmonic components in a signal can be detected regardless of the phase. Note that this feature cannot be achieved by real wavelets such as the Meyer wavelets, which are all self-similar, being derived from a unique mother wavelet by scaling and shifting. Also, note that orthonormal basis of real wavelets can be generated by considering either the real or the imaginary parts only of Equation 6.48. For instance, the well-known Shannon wavelets correspond to the imaginary parts of Equation 6.48, for $m, n = 1, 2; 2, 4; 4, 8; \ldots$. Harmonic wavelets are used in many mechanics applications such as acoustics, vibration monitoring, and damage detection [45–49].

6. *Complex Gaussian wavelets* — Families of wavelets, each corresponding to a pth order derivative of a complex Gaussian function. That is,

$$\psi_p(x) = C_p \frac{d^p}{dx^p} (e^{-ix} e^{-x^2/2}), \quad p = 1, 2, \ldots \tag{6.49}$$

where C_p is a normalization constant such that $\|\psi(x)\|^2 = 1$. Complex Gaussian wavelets support the CWT only. They have no finite support in the x-domain, although the interval $[-5, 5]$ is generally taken as effective support. Despite their lack of orthogonality, they are quite popular in image-processing applications due to their regularity [1].

7. *Complex Morlet wavelets* — Families of [50], each obtained as the derivative of the classical Morlet wavelet $\psi_0(x) = e^{-x^2/2} e^{i\omega_0 x}$. That is,

$$\psi_p(x) = \frac{d^p}{dx^p} (e^{-x^2/2} e^{i\omega_0 x}), \quad p = 1, 2, \ldots \tag{6.50}$$

Except for $\psi_0(x)$, which does not satisfy the admissibility condition (Equation 6.11) in a strict sense, all the other members of the family are proper wavelets. For practical purposes, however, $\psi_0(x)$ is generally considered admissible for $\omega_0 \geq 5$. Complex Morlet wavelets support the CWT only and are not orthogonal. However, they are all progressive, that is, they satisfy the condition posed by Equation 6.36. Further, for the Morlet wavelet $\psi_0(x)$, there exists a relation between the scale parameter a and the frequency ω at which its Fourier transform focuses. That is,

$$a = \frac{\omega_0}{\omega} \tag{6.51}$$

Complex Morlet wavelets are then applied for structural identification purposes, as shown in Section 6.5.

6.3 Time-Dependent Spectra Estimation of Stochastic Processes

Wavelets-based approaches are significant tools for joint time–frequency analysis of problems related to vibrations of mechanical and structural systems. This applies both to the characterization of the system excitation, the system identification, and the system response determination. Several examples exist in nature of stochastic phenomena with a time-dependent frequency content. The frequency content of earthquake records, for instance, evolves in time due to the dispersion of the propagating seismic waves [51, 52]. Further, sudden changes in the wave frequency at a given location of the sea surface are often induced by fast-moving meteorological fronts [53]. Also, a rapid change in the frequency content is generally associated with waves at the breaking stage. Similarly, turbulent gusts of time-varying frequency content are often embedded in wind fields.

Appropriate description of such phenomena is obviously crucial for design and reliability assessments. In an early attempt, concepts of traditional Fourier spectral theory were generalized to provide spectral estimates, such as the Wigner–Ville method [25, 54] or the CGT of Equation 6.1. However, it soon became clear that the extension of the traditional concept of a spectrum is not unique, and proposed time-varying spectra could have contradictory properties [6, 55].

Wavelet analysis is readily applicable for estimating time-varying spectral properties, and a significant effort has been devoted to formulating "wavelet energy principles" that work as alternatives to classical Fourier methods. Measures of a time-varying frequency content were first obtained by "sectioning," at different time instants, the wavelet coefficients mean square map [49, 56–58]. Developing consistent spectral estimates from such sections, however, is not straightforward. From a theoretical point of view, it either requires an appropriate wavelet-based definition of time-varying spectra, or it must relate to well-established notions of time-varying spectra. From a numerical point of view, it involves certain difficulties in converting the scale axis to a frequency axis, especially when the wavelet functions are not orthogonal in the frequency domain; that is, when the frequency content of wavelet functions at adjacent scales do overlap.

Early investigations on wavelet-based spectral estimates may be found in references such as [44, 59–64], where wavelet analysis was applied in the context of earthquake engineering problems. In a particular approach, a modified Littlewood Paley (MLP) wavelet basis can be introduced, whose mother wavelet is defined in the frequency domain by the equation

$$\hat{\Psi}(\omega) = \begin{cases} \dfrac{1}{\sqrt{2(\sigma - 1)\pi}}, & \pi \le |\omega| \le \sigma\pi, \\ 0, & \text{elsewhere} \end{cases} \tag{6.52}$$

In Equation 6.52, the symbol σ denotes a scalar factor, to be adjusted depending on the desired frequency resolution. The MLP wavelets are orthogonal in the frequency domain, that is, wavelets at adjacent scales span nonoverlapping intervals. The MLP wavelets have been used in conjunction with a discretized version of the CWT proposed by Alkemade [65] for a finite-energy process $f(t)$

$$f(t) = \sum_{i,j} \frac{K\Delta b}{a_j} W_f(a_j, b_i)\psi_{a_j, b_i}(t) \tag{6.53}$$

where $a_j = \sigma^j$, Δb is a time step, and K is a constant parameter depending on σ.

In many instances, Equation 6.53 can be construed as representing realizations of a *stochastic process*, and in this case, the following estimate of its instantaneous mean-square value of $f(t)$ has been constructed

$$E[f^2(t)]\big|_{t=b_i} = K \sum_j \frac{E[W_f(a_j, b_i)]^2}{a_j} \tag{6.54}$$

where $E[\cdot]$ is the mathematical expectation operator over the ensemble of realizations. From Equation 6.54, and based on the orthogonality properties of the MLP wavelets, the following quantity

$$S_f(\omega)\big|_{t=b_i} = \sum_j K \frac{E[W_f(a_j, b_i)]^2}{a_j} \left|\hat{\Psi}_{a_j, b_i}(\omega)\right|^2 \tag{6.55}$$

where the symbol $\hat{\Psi}_{a_j, b_i}(\omega)$ denotes the Fourier transform of the wavelet function $\psi_{a_j, b_i}(t)$, can be taken as a measure of the time-varying power spectral density (PSD) of the process $f(t)$. Based on Equation 6.55, closed-form expressions can be derived between the input and the output PSDs [63]. In this context, linear-response statistics, such as the instantaneous rate of crossings of the zero level or the instantaneous rate of occurrence of the peaks, have been estimated with considerable accuracy. Analysis of nonlinear systems has also been attempted by an equivalent statistical linearization procedure [61, 66].

Wavelet analysis for spectral estimation has also been pursued by Kareem et al., who have used the squared wavelet coefficients of a DWT to estimate the PSD of stationary processes [56]. To improve the frequency resolution of the DWT, where only adjacent octave bands can be accounted for, a CWT can be implemented based on a complex Morlet wavelet basis. The latter is preferable due to the one-to-one correspondence between the scale a and the center frequency (Equation 6.51), which allows minimizing the overlap between spectral estimates at adjacent scales. Further, the product of wavelet coefficients can be used as a measure of the cross-correlation between two nonstationary signals $x(t)$ and $y(t)$ [56]. This concept can be refined by the introduction of a wavelet *coherence* measure [57, 58] expressed by the equation

$$(c^{\mathrm{W}}(a,b))^2 = \frac{\left|S_{xy}^{\mathrm{W}}(a,b)\right|^2}{S_{xx}^{\mathrm{W}}(a,b)S_{yy}^{\mathrm{W}}(a,b)} \qquad (6.56)$$

In this equation, the local spectrum $S_{ij}^{\mathrm{W}}(a,b)$ is defined as

$$S_{ij}^{\mathrm{W}}(a,b) = \int_T \overline{W}_i(a,b)W_j(a,b)\mathrm{d}\tau \qquad (6.57)$$

where the time integration window T depends on the desired time resolution. The local spectrum (Equation 6.57), owing to the time average over T, allows smoothing of potential measurement noise effects. Measures of higher-order correlation can also be introduced [56, 58], such as the wavelet *bicoherence*

$$(b_{xxy}^{\mathrm{W}}(a_1,a_2,b))^2 = \frac{\left|B_{xxy}^{\mathrm{W}}(a_1,a_2,b)\right|^2}{\int_T \left|W_x(a_1,\tau)W_x(a_2,\tau)\right|^2 \mathrm{d}\tau \int_T \left|W_y(\tilde{a},\tau)\right|^2 \mathrm{d}\tau} \qquad (6.58)$$

where $1/\tilde{a} = 1/a_1 + 1/a_2$, and

$$B_{xxy}^{\mathrm{W}}(a_1,a_2,b) = \int_T W_x(a_1,\tau)W_x(a_2,\tau)W_y(\tilde{a},\tau)\mathrm{d}\tau \qquad (6.59)$$

Related remedies can be adopted to suppress spurious correlations induced by statistical noise, based on a reference noise map created from artificially simulated signals [58].

Signal energy representation concepts have been examined in Ref. [67] by using quasi-orthogonal Daubechies wavelets in the frequency domain to simulate earthquake ground motion accelerations. Further, Massel has used wavelet analysis to capture time-varying frequency composition of sea-surface records due to fast-moving atmospheric fronts in deep water, wave growth, and breaking or disintegration of mechanically generated wave trains [68]. In this regard, absolute value wavelet maps and a spectral measure called global wavelet energy spectrum, defined by the equation

$$E_3(a) = \int_0^\infty E_1(a,b)\mathrm{d}b \qquad (6.60)$$

are used. The symbol $E_1(\tau,b)$ denotes a time-scale energy density

$$E_1(a,b) = \frac{\left|W_{\mathrm{f}}(a,b)\right|^2}{a} \qquad (6.61)$$

The scale in Equation 6.61 is readily translated into frequency by selecting the Morlet wavelet basis.

Spanos and Failla [69] have applied wavelet analysis to estimate the evolutionary power spectral density (EPSD) of nonstationary oscillatory processes defined as [70]

$$f(t) = \int_{-\infty}^\infty A(\omega,t)\mathrm{e}^{\mathrm{i}\omega t}\,\mathrm{d}Z(\omega) \qquad (6.62)$$

The symbol $A(\omega,t)$ denotes a slowly varying time- and frequency-dependent modulating function, and $Z(\omega)$ is a complex random process with orthogonal increments such that $E\big[|\mathrm{d}Z(\omega)|^2\big] = S_{f_0f_0}(\omega)\mathrm{d}\omega$, where $S_{f_0f_0}(\omega)$ is the two-sided PSD of the zero-mean stationary process

$$f_0(t) = \int_{-\infty}^\infty \mathrm{e}^{\mathrm{i}\omega t}\,\mathrm{d}Z(\omega) \qquad (6.63)$$

The two-sided EPSD of $f(t)$ is then taken as

$$S_{ff}(\omega, t) = \left|A(\omega, t)\right|^2 S_{f_0 f_0}(\omega) \tag{6.64}$$

Due to its localization properties, the wavelet transform of $f(t)$ (Equation 6.62) may be approximated as an oscillatory stochastic process. That is,

$$W_f(a, b) \approx \int_{-\infty}^{\infty} A(\omega, b)e^{i\omega b}\, dZ'(\omega) \tag{6.65}$$

where $dZ'(\omega) = \sqrt{2\pi a}\,\overline{\hat{\Psi}}(\omega a)dZ(\omega)$. Based on Equation 6.65, the following integral relation is found between the mean-squared wavelet coefficients at each scale a and the EPSD of $f(t)$. That is,

$$E[W_f(a, b)^2] = 4\pi a \int_0^{\infty} \left|\hat{\Psi}(\omega a)\right|^2 S_{ff}(\omega, b)d\omega \tag{6.66}$$

A sufficient number of Equation 6.66, one for each scale a, can be solved by a standard solution algorithm, applicable for both orthogonal and nonorthogonal bases in the frequency domain. This procedure has proved quite accurate using both the Littlewood–Paley and the real Morlet wavelet bases.

6.4   Random Field Simulation

The use of wavelets for random field synthesis can be examined within the more general framework of scale-type methods. The latter have been developed to improve the computational performances of Monte Carlo simulations. Classical methods such as the spectral approach [71] or the autoregressive moving average (ARMA) [72] are not readily applicable for this purpose, especially when using nonuniform meshes or when enhancement of local resolution is desirable. To address these shortcomings, Fournier et al. [73] have proposed a "random midpoint method" to synthesize fractional Brownian motion; that is, a scale-type method where values of the random field for points within a coarser scale are generated first, and then the generated samples are used to determine values for a finer scale. This approach has been extended by Lewis [74] into a "generalized stochastic subdivision method," suitable for a broad class of stationary processes, and by Fenton and Vanmarcke [75] into a "local average subdivision method," which includes a random field smoothing procedure producing averages of the field for an increasingly finer scale.

An interpretation of scale-type approaches in the context of random field synthesis has been given by Zeldin and Spanos [39] using compactly supported Daubechies wavelets. Specifically, a synthesis algorithm has been developed that includes the previous methods proposed by Lewis [74] and Fenton and Vanmarcke [75] as a particular case. To synthesize a sample of a given process, the closed-form expressions

$$r_{k,l}^{j,i} = E\left[d_k^j d_l^i\right] = \int_{-\infty}^{\infty} \int_{-\infty}^{\infty} R_f(x_1, x_2)\psi_{j,k}(x_1)\psi_{i,l}(x_2)dx_1\, dx_2 \tag{6.67}$$

$$b_{k,l}^{j,i} = E\left[c_k^j d_l^i\right] = \int_{-\infty}^{\infty} \int_{-\infty}^{\infty} R_f(x_1, x_2)\phi_{j,k}(x_1)\psi_{i,l}(x_2)dx_1\, dx_2 \tag{6.68}$$

$$a_{k,l}^{j,i} = E\left[c_k^j c_l^i\right] = \int_{-\infty}^{\infty} \int_{-\infty}^{\infty} R_f(x_1, x_2)\phi_{j,k}(x_1)\phi_{i,l}(x_2)dx_1\, dx_2 \tag{6.69}$$

given in Refs. [21, 39] are considered to relate the autocorrelation function $R_f(x_1, x_2)$ of the process to the coefficients of its wavelet transform, which in this case are random variables. The synthesis algorithm is based on the wavelet reconstruction algorithm developed by Mallat [34, 35], which proceeds from coarse to fine scales to determine the wavelet coefficients. Some relevant properties of wavelet ensure the computational efficiency of the algorithm. Specifically, using the quasi-differential properties of wavelets showed by Belkin [76], the coefficients d_k^j's are derived directly from c_k^j's by the approximate linear combination

$$d_k^j = \sum_l \alpha_{k,l}^j c_l^j + \beta_k^j u_k \tag{6.70}$$

where u_k's are uncorrelated, zero mean, unit variance random variables, statistically independent of c_k^j's. For a wide class of stochastic processes, wavelet coefficients prove weakly correlated as the difference $k - l$ increases and, for this, the summation in Equation 6.70 is generally restricted to adjacent elements only. The algorithm is completed by an error-assessment procedure which allows refining of the triggering scale j in order to fit the sought target statistical properties of the synthesized field.

Further studies on the role of wavelet analysis in stochastic mechanics applications may be found in Ref. [21], which has showed how wavelet bases can be used in approximate Karhunen–Loève expansions. Any stationary process can then be represented as

$$f(t) = \sum_{j,k} d_k^j \psi_{j,k}(t) \tag{6.71}$$

where d_k^j's are uncorrelated random variables and $\psi_{j,k}(t)$ are a nonorthogonal wavelet-like basis.

6.5   System Identification

Wavelet analysis lends itself to system-identification applications. For instance, frequency localization properties allow detection and decoupling of individual vibration modes of multi-degree-of-freedom (multi-DoF) linear systems. The wavelet representation of the system response can be truncated to an appropriate scale parameter in order to filter measurement noise. Also, the wavelet transform coefficients can be related directly to the system parameters, as long as specific bases are used.

Early investigations trace back to the work by Robertson et al. [77], who have used the DWT for the estimation of the impulse-response function of multi-DoF systems. Compared with alternative time-domain techniques, the DWT-based extraction procedure offers significant advantages. It is robust since singularities in the procedure-related matrices can generally be avoided by selecting orthonormal wavelet functions. Further, the reconstructed impulse-response function captures the low-frequency components, referred to as static modes and mode shape errors, which ordinarily are difficult to estimate.

An important application of wavelet analysis to structural identification is due to Staszewski [78], who has used complex Morlet wavelets for modal damping estimation. Specifically, Staszewski has interpreted in terms of the wavelet transform some concepts already used in well-established methods, where the Hilbert transform has been applied to a free-vibration linear response [79]. In the case of light damping, the free response in each mode $x_j(t)$ may be approximated in the complex plane by an analytical signal, given by Equation 6.37. The modulus of the Morlet wavelet transform of $x_j(t)$ can be expressed as

$$\left|W_{x_j}(a,b)\right| \approx A_j\, e^{-\zeta_j \omega_j b} \left|\bar{\hat{\Psi}}(\pm ia_j \omega_j \sqrt{1 - \zeta_j^2})\right| \tag{6.72}$$

where A_j is the residue magnitude, and ω_j and ζ_j are the mode natural frequency and damping ratio, respectively. In Equation 6.72, the symbol a_j denotes the specific scale value, related to the mode natural frequency ω_j by the closed-form relation 6.51, typical of Morlet wavelets. Assuming that the natural frequency ω_j has been previously computed, the damping ratio ζ_j can then be estimated as the slope of a straight line, representing the cross section wavelet modulus (Equation 6.72) plotted in a semilogarithmic scale. That is,

$$\ln \left|(W_{x_j}(a_j, b))\right| \approx -\zeta_j \omega_j b + \ln\left(A_j \left|\bar{\hat{\Psi}}(\pm ia_j \omega_j \sqrt{1 - \zeta_j^2})\right|\right) \tag{6.73}$$

Staszewski has also proposed an alternative damping estimation method based on the *ridge* and *skeletons* of the wavelet transform. A ridge is a curve of local maxima in the mean-square wavelet map, and the corresponding skeleton is given by the values of the wavelet transform restricted to the ridge. As a result of the localization properties of the wavelet transform, the ridges and skeletons of the wavelet transform can be detected separately for each mode. Specifically, the real part of the skeleton of the wavelet transform gives the impulse-response function for each single mode from which a straightforward estimate of the damping ratio is obtained from a logarithmic equation analogous to Equation 6.73. A generalization of the method for nonlinear systems can also be formulated [80].

Ruzzene et al. [81] have also presented a damping estimation algorithm based on the same concepts and leading to analogous results. Certain issues have been addressed in detail concerning the frequency resolution of the adopted wavelet basis, crucial for detecting coupled modes, and appropriate algorithms for ridges extraction [50]. Lardies and Gouttebroze [82] have estimated modal parameters *via* ambient records without input measurements. To this end, the random decrement method (see Ref. [83] and the references therein) has been used to convert ambient vibration response into a free vibration response. Also, a modified Morlet wavelet basis has been developed with enhanced properties for modal parameters estimation. The method devised by Staszewski and by Ruzzene et al. has also been implemented by Slavic et al. [84] by replacing the Morlet wavelets by Gabor wavelets, whose time and frequency resolution may be adjusted by an appropriate parameter. Explicit conditions have been given on the frequency *bandwidths* of the Gabor wavelet transform, in order to estimate the instantaneous frequencies of two adjacent modes.

Damping coefficients have been estimated using a logarithmic decrement formula, where the ratio of the wavelet transform at two subsequent extremes of the pseudo-period $T_j = 2\pi/\omega_j$ of the response in each mode is involved for a selected wavelet transform scale [85, 86]. For the procedure to estimate the damping coefficient associated with the fundamental mode, it is sufficient to adapt the analyzing scale so that the higher frequency modes are filtered. For an arbitrary mode j, low-pass filtering is used to cancel the fundamental and the first $j-1$ modes. Ghanem and Romeo [87] have formulated a wavelet-Galerkin method for time-varying systems, where both damping and stiffness parameters are computed by solving a matrix equation. The latter is built by a standard Galerkin method by projecting the solution of the differential equation of motion onto a subspace described by the wavelet scaling functions of a compactly supported Daubechies wavelet basis. The method is accurate for both free and forced vibration responses. A formulation for nonlinear systems has also been proposed [88]. Another application is due to Yu and Xiao [89], who have used wavelet transform to identify the parameters of a Preisach model of hysteresis (see Refs. [69, 90] and the references therein). The output function of the Preisach model is expanded in terms of the scaling functions of a given wavelet basis. Then, the coefficients of such an expansion are determined by fitting a number of experimental data points with a minimum energy method. From the output function, the so-called Preisach function can be determined in a closed form.

A comprehensive application of wavelet-analysis concepts to system-identification problems has been given by Le and Argoul [91]. They have developed closed-form expressions to compute the damping ratio, the natural frequency and the shape of each mode, based on ridges and skeletons of the wavelet transformed free vibration response. As an alternative, Yin et al. [92] have proposed to apply the wavelet transform to the frequency response function (FRF) of the system. Specifically, given the FRF of an N-DoF system in the form

$$H(\omega) = \sum_{r=1}^{N} \left[\frac{A_r}{i\omega - \lambda_r} + \frac{\overline{A}_r}{i\omega - \overline{\lambda}_r} \right] \tag{6.74}$$

where λ_r is the rth complex pole and A_r the rth residue, a complex fractional function

$$\psi_y(x) = \frac{1}{(1 + ix)^{y+1}} = e^{-(y+1)\log(1+ix)}, \quad y \in \mathbb{R}^+ \tag{6.75}$$

is selected as a wavelet basis. Based on Equation 6.75, a closed-form expression can be established for the CWT of Equation 6.74 multiplied by $(\sqrt{a})^{-y}$. Specifically,

$$H_y(a, b) = a^{-(y+1)/2} \int_{-\infty}^{\infty} H(\omega) \bar{\psi}_y \left(\frac{\omega - b}{a} \right) d\omega$$

$$= 2\pi a^{(y+1)/2} \sum_{r=1}^{N} \left(\frac{A_r}{(a + ib - \lambda_r)^{y+1}} + \frac{\bar{A}_r}{(a + ib - \bar{\lambda}_r)^{y+1}} \right) \tag{6.76}$$

Natural frequencies and damping ratios can be estimated by locating the maxima of Equation 6.76 in the (a, b) plane.

6.6 Damage Detection

Properties of the wavelet transform are also quite appealing for damage-detection purposes. Early investigations in this field [93, 94] used wavelet analysis to detect local faults in machineries. Specifically, visual inspection of the modulus and phase of the wavelet transform was used to localize the fault [93]. Further, it was shown that transient vibrations due to developing damage are disclosed by the local maxima of the mean-square wavelet map [94]. These investigations gave a qualitative approach to damage detection as no estimate of the damage amplitude was provided. Additional studies have confirmed the correlation between local maxima of the wavelet transform and damage in beams and plates [95–97], and a first attempt to estimate the damage amplitude was made by Okafor and Dutta [98]. Specifically, Daubechies wavelets were used to wavelet transform the mode shapes of a damaged cantilever beam, and a regression analysis by a least-square method was conducted to correlate the peaks of the wavelet coefficients with the corresponding damage amplitude.

A consistent mathematical framework for wavelet analysis of damaged beams is due to Hong et al. [99]. The focal concept is that defects in structures, even if small, may affect significantly the vibration mode shapes, depending on the location and the kind of damage. Such variations may not be apparent in the measured data but become detectable as singularities if wavelet analysis is used due to its high resolution properties. Specifically, Hong et al. have shown that the singularity of the vibration modes can be described in terms of Lipschitz regularity, a concept also encountered in the theory of differential equations, widely used in image processing where object contours correspond to irregularities in the intensity [100, 101]. In mathematical terms, a function $f(x)$ is Lipschitz $\alpha \geq 0$ at $x = x_0$ if there exists $K > 0$, and a polynomial of order m (m is the largest integer satisfying $m \leq \alpha$), $p_m(x)$, such that and a polynomial of order m, $p_m(x)$, such that

$$f(x) = p_m(x) + \varepsilon(x) \tag{6.77}$$

$$\left| \varepsilon(x) \right| \leq K \left| x - x_0 \right|^{\alpha} \tag{6.78}$$

The wavelet transform of Lipschitz α functions enjoys some properties. Mallat and Hwang [100] have shown that for a wavelet basis with a number of vanishing moments $\alpha \leq n$, a local Lipschitz singularity at x_0 corresponds to maxima lines of the wavelet transform modulus. That is, local maxima with asymptotic decay across scales. Near the cone of influence $x = x_0$, such moduli satisfy the equation

$$\left| W_f(a, x) \right| \leq A a^{\alpha + 1/2}, \quad A > 0 \tag{6.79}$$

from which the Lipschitz exponent is computed as

$$\log_2 \left| W_f(a, x) \right| \leq \log_2 A + \left(\alpha + \frac{1}{2} \right) \log_2 a \tag{6.80}$$

By plotting the wavelet coefficients on a logarithmic scale, A and α may be computed by setting the equality sign in Equation 6.80 and minimizing the error in the least-square sense. Hong et al. have applied Equation 6.79 to the first mode shape of a damaged cantilever beam *via* a Mexican Hat wavelet transform. The first mode shape is preferable since it is the most accurately determined by modal testing; it features the lowest curvature; and sets off the singularity better. A correlation between damage size and the magnitude of the Lipschitz exponent has been found from a number of beams with different damage parameters.

Some of the ideas presented by Hong et al. may also be found in the work by Douka et al., who have pursued crack identification in beams and plates using Daubechies wavelets [102, 103]. The first mode vibration response has been considered and the singularity induced by local defects has been characterized in terms of Equation 6.79. The Lipschitz exponent has been used to describe the kind of singularity, and the parameter A has been taken as the factor relating the depth of the crack to the amplitude of the wavelet transform. Specifically, a second-order polynomial law has been found for the intensity factor as a function of the crack depth. The work by Douka et al. has pointed out the importance of the number of vanishing moments M of the chosen wavelet basis. It is intuitive that the capability of setting off singularities in a regular function increases with M. However, wavelet functions with high M exhibit a long support and lack space resolution. A compromise, then, must be achieved, depending on the application in hand. Further insight into some mathematical details of both the methods developed by Hong et al. and Douka et al. may be found in Haase and Widjajakusuma [50]. Specifically, a fast algorithm to determine the maxima lines of the wavelet transform has been devised. Also, the performance of various wavelet bases, such as the Gaussian family of wavelets, has been assessed versus Daubechies wavelets used by Douka et al.

Another approach for damage-detection problems has been proposed by Yam et al. [104]. Clearly, detection of small and incipient damage cannot be pursued by computing modal parameters that change only if the amount of damage is significant. Thus, a method has been devised based on the energy variation of the vibration response due to the occurrence of damage. The method is implemented in two steps. The first involves the construction of damage feature proxy vectors using the energy at various scales of the wavelet transformed vibration response. Then, classification and identification of the structural damage status is pursued by using artificial neural networks (ANNs), which offer significant advantages compared with genetic algorithms (GAs) developed by Moslem and Nafaspour for damage-identification purposes [105]. Genetic-algorithm-based damage detection, in fact, requires repeatedly searching among numerous damage parameters to find the optimal solution of the objective function.

Yet another approach for applications of wavelet analysis to damage detection has been discussed by Paget et al. [106], who have developed a procedure to detect impact damage in composite plates. It is based on Lamb waves generated and received by embedded piezoceramic transducers. The Lamb waves can be quite effective since they can propagate over long distances in the composite material and can interfere with damage. To characterize the damage, the Lamb waves are wavelet transformed using an original wavelet basis, devised from the recurrent waveforms of the Lamb waves. The changes in the Lamb waves interacting due to the occurrence of damage are captured by the amplitude change of the wavelet coefficients. From this effect, an estimate of the impact energy and the damage level is obtained based on experimental results.

6.7 Material Characterization

Material properties description is another application for wavelet analysis. Intuition suggests that multiscale analysis is a natural way of describing microstructure or material heterogeneity. Various, in fact, are the examples of multiscale microstructures, such as porosity distributions in ceramics, defects, dislocations, grain boundaries, and pores. It is important, however, to understand how information at different scales is related, and whether large or small scales affect macroscopic material properties such as deformation, toughness, and electrical conductance. Further interest towards a multiscale description of material properties is motivated by the need of alternatives to the standard finite element method (FEM).

The latter, although capable in principle, cannot simulate the actual behavior of materials such as aluminum alloys, where pores may attain a size up to 500 μm and inclusions may attain a size of 3 to 6 μm in diameter. Further, in FEM-based methods, the constitutive response of the material at increasing scales is not the result of microstructural analysis at smaller scales, but it is rather assumed on the basis of macroscopic experiments.

Willam et al. [107] have performed *multiresolution* homogenization based on a recursive Schur reduction method in conjunction with the Haar wavelet transform. The method allows coarse-grained parameters, such as Young's modulus of elasticity, to be extracted from fine-grained properties at the meso- and microscales. Also, progressive elastic degradation can be modeled, which initiates at a quite fine scale and evolves into a macroscopic zero stiffness at the continuum level.

Frantziskonis [108] has focused on stationary and isotropic porous media. The geometry of porous media is generally described in terms of a fundamental function, defined as unity for spatial locations in the matrix, and as zero for locations in the pores or flaws. At a solid-flaw interface, the porous medium is represented mathematically through a local jump in the fundamental function. It has been found that such a jump can be captured by a wavelet transform, as long as the finest scale is small enough relative to the size of the pores. From this fact, a relationship between the energy of the wavelet transform of the porous medium, and the variance and the correlation distance of the solid phase can be derived. In the presence of heterogeneous materials, with multiscale porosity, the role of porosity at each scale has been identified through the variation of the energy of the wavelet transform as a function of scale. Peaks of the energy reveal the dominant scale in determining macroscopic properties of the materials, such as mechanical failure. Specifically, a biorthogonal spline with four vanishing moments has been employed as a wavelet basis. The results obtained have been subsequently extended in a second study, addressing the crack formation in an aluminum alloy with distributed pores and inclusions [109]. The problem, implemented for a one-dimensional solid, is tackled by wavelet transforming the flexibility function, assumed to vary along the longitudinal axis of the one-dimensional solid. The relationship between the energy of the wavelet transform and the variance of the flexibility is used to detect the dominant scale in the crack-formation process.

Note that an application of a two-dimensional wavelet transform has been described in Ciliberto et al. [110] for porosity classification on carbon fiber-reinforced plastics.

6.8 Concluding Remarks

Concepts of wavelets-based continuous and discrete representations have been reviewed. Further, an overview of vibration-related applications for evolutionary spectrum estimation, random field simulation, system identification, damage detection, and material characterization has been included. The list of references is not exhaustive. However, these references can serve as readily available resources for canvassing the multitude of concepts and applications of this remarkable tool for capturing and representing localization features of many physical phenomena. Wavelets-based algorithms and commercial codes are an indispensable family of tools of vibration analysis and offer, in many cases, a potent improvement over the classical Fourier-transform-based approaches.

Acknowledgments

The support of this work through a grant from the U.S. Department of Energy is gratefully acknowledged.

Nomenclature

Symbol	Quantity	Symbol	Quantity
$E[\cdot]$	the operator of mathematical expectation	ω	frequency
$R(\cdot,\cdot)$	correlation function	ζ	damping ratio
$H[\cdot]$	Hilbert transform operator	a	scale
$\langle\cdot,\cdot\rangle$	inner product	b	shift
$\subset$	inclusion	$\psi(x)$	mother wavelet
$\{\cdot;\cdot\}$	set of all elements with a specified property	$\phi(x)$	scale function
		x	spatial variable
$\lvert\cdot\rvert$	absolute value	W	wavelet transform
$\lVert\cdot\rVert$	norm	i	$\sqrt{-1}$
$\mathbb{Z}$	the set of integer numbers	δ_{mn}	Kronecker delta defined as
$\mathbb{C}$	the set of complex numbers		$\delta_{mn} = \begin{cases} 0 & \text{for } m \neq n \\ 1 & \text{for } m = n \end{cases}$
$\overline{(\cdot)}$	complex conjugate		

References

1. Carmona, R., Hwang, W.-L., and Torrésani, B. 1998. *Practical Time–Frequency Analysis: Gabor and Wavelet Transforms with an Implementation in S.*, Academic Press, San Diego.
2. Chan, Y.T. 1995. *Wavelet Basics*, Kluwer, Boston.
3. Chui, C.K. 1992. *An Introduction to Wavelets*, Academic Press, New York.
4. Chui, C.K. 1992. *Wavelets: A Tutorial in Theory and Applications*, Academic Press, New York.
5. Chui, C.K., Montefusco, L., and Puccio, L. 1994. *Wavelets: Theory, Algorithms, and Applications*, Academic Press, New York.
6. Cohen, L. 1995. *Time–Frequency Analysis*, Prentice Hall, Englewood Cliffs, NJ.
7. Daubechies, I. 1992. *Ten Lectures on Wavelets*, Society for Industrial and Applied Mathematics, Philadelphia.
8. Hernández, E. and Weiss, G.L. 1996. *A First Course on Wavelets*, CRC Press, Boca Raton, FL.
9. Hubbard, B.B. 1996. *The World According to Wavelets: The Story of a Mathematical Technique in the Making*, A.K. Peters, Wellesley, MA.
10. Jaffard, S., Meyer, Y., and Ryan, R.D. 2001. *Wavelets: Tools for Science & Technology*, Society for Industrial and Applied Mathematics, Philadelphia.
11. Kahane, J.-P. and Lemarié-Rieusset, P.-G. 1995. *Fourier Series and Wavelets*, Gordon & Breach, Luxembourg.
12. Kaiser, G. 1994. *A Friendly Guide to Wavelets*, Birkhäuser, Boston.
13. Mallat, S.G. 1998. *A Wavelet Tour of Signal Processing*, Academic Press, San Diego.
14. Meyer, Y. 1990. *Ondelettes*, Hermann, Paris.
15. Misiti, M. 1997. *Wavelet Toolbox for Use with Matlab*, Math Works, Natick, MA.
16. Newland, D.E. 1993. *An Introduction to Random Vibrations, Spectral and Wavelet Analysis*, 3rd ed., Longman Scientific & Technical, New York, Harlow, UK.
17. Qian, S. and Chen, D. 1996. *Joint Time–Frequency Analysis: Methods and Applications*, Prentice Hall, Upper Saddle River, NJ.
18. Qian, S. 2001. *Introduction to Time–Frequency and Wavelet Transforms*, Prentice Hall, Upper Saddle River, NJ.
19. Strang, G. and Nguyen, T. 1996. *Wavelets and Filter Banks*, Cambridge Press, Wellesley, MA.
20. Vetterli, M. and Kovacevic, J. 1995. *Wavelets and Subband Coding*, Prentice Hall, Englewood Cliffs, NJ.

21. Walter, G.G. 1994. *Wavelets and Other Orthogonal Systems with Applications*, CRC Press, Boca Raton, FL.
22. Young, R.K. 1993. *Wavelet Theory and Its Applications*, Kluwer, Boston.
23. Spanos, P.D. and Zeldin, B.A., A state-of-the-art report on computational stochastic mechanics: wavelets concepts, *Probab. Eng. Mech.*, 12, 244, 1997.
24. Gabor, D., Theory of communication, *J. Inst. Electr. Eng.*, 93, 429, 1946.
25. Wigner, E.P., On the quantum correction for thermodynamic equilibrium, *Phys. Rev.*, 40, 749, 1932.
26. Goupillaud, P., Grossmann, A., and Morlet, J., Cycle-octave and related transforms in seismic signal analysis, *Geoexploration*, 23, 85, 1984.
27. Grossmann, A. and Morlet, A., Decomposition of hardy functions into square integrable wavelets of constant shape, *SIAM J. Math. Anal.*, 15, 723, 1984.
28. Holschneider, M. and Tchamitchian, P., Pointwise analysis of Riemann's 'nondifferentiable' function, *Invent. Math.*, 105, 157, 1991.
29. Cohen, A. and Kovacevic, J., Wavelets: the mathematical background, *Proc. IEEE*, 84, 514, 1996.
30. Duffin, R.J. and Schaeffer, A.C., A class of nonharmonic Fourier series, *Trans. Am. Math. Soc.*, 72, 341, 1952.
31. Daubechies, I., The wavelet transform, time–frequency localization and signal analysis, *IEEE Trans. Inf. Theory*, 36, 961, 1990.
32. Daubechies, I., Grossmann, A., and Meyer, Y., Painless nonorthogonal expansions, *J. Math. Phys.*, 27, 1271, 1986.
33. Battle, G., A block spin construction of ondelettes. I. Lemarie functions, *Commun. Math. Phys.*, 110, 601, 1987.
34. Mallat, S.G., Multifrequency channel decompositions of images and wavelet models, *IEEE Trans. Acoust. Speech Signal Process.*, 37, 2091, 1989.
35. Mallat, S.G., A theory for multiresolution signal decomposition: the wavelet representation, *IEEE Trans. Pattern Anal. Machine Intell.*, 11, 674, 1989.
36. Daubechies, I., Orthogonal bases of compactly supported wavelets, *Commun. Pure Appl. Math.*, 41, 909, 1988.
37. Basseville, M., Benveniste, A., Chou, K.C., Golden, S.A., Nikoukhah, R., and Willsky, A.S., Modeling and estimation of multiresolution stochastic processes, *IEEE Trans. Inf. Theory*, 38, 766, 1992.
38. Basseville, M., Benveniste, A., and Willsky, A.S., Multiscale autoregressive processes. Part I: Schur–Levinson parametrizations, *IEEE Trans. Signal Process.*, 40, 1915, 1992.
39. Zeldin, B.A. and Spanos, P.D., Random field representation and synthesis using wavelet bases, *ASME J. Appl. Mech.*, 63, 946, 1996.
40. Carmona, R.A., Hwang, W.L., and Torresani, B., Characterization of signals by the ridges of their wavelet transforms, *IEEE Trans. Signal Process.*, 45, 2586, 1997.
41. Kim, H.O., Kim, R.Y., and Lim, J.K., Characterizations of biorthogonal wavelets which are associated with biorthogonal multiresolution analyses, *Appl. Comput. Harmonic Anal.*, 11, 263, 2001.
42. Spanos, P.D. and Rao, V.R.S., Random field representation in a biorthogonal wavelet basis, *J. Eng. Mech.*, 127, 194, 2001.
43. Newland, D.E., Wavelet analysis of vibration, Part I: Theory, *J. Vib. Acoust.*, 116, 409, 1994.
44. Spanos, P.D., Tratskas, P.N., and Tezcan, J., Stochastic processes evolutionary spectrum estimation via the wavelet spectrum, *J. Comput. Methods Appl. Mech. Eng.*, 194, 12–16, 1367–1383, 2005.
45. Newland, D.E., Wavelet analysis of vibration, Part II: Wavelet maps, *J. Vib. Acoust.*, 116, 417, 1994.
46. Newland, D.E., Some properties of discrete wavelet maps, *Probab. Eng. Mech.*, 9, 59, 1994.
47. Newland, D.E. and Butler, G.D. 1998. Application of time–frequency analysis to strong motion data with damage, *Proceedings of the 69th Shock and Vibration Symposium*, Minneapolis, St Paul, MN.

48. Newland, D.E., Ridge and phase identification in the frequency analysis of transient signals by harmonic wavelets, *J. Vib. Acoust.*, 121, 149, 1999.

49. Newland, D.E. and Butler, G.D. 1999. Time-varying cross-spectra by harmonic wavelets for soil motion damage, *Proceedings of the ASME DETC: 17th ASME Biennial Conference on Mechanical Vibration and Noise*, Las Vegas, NV, p. 1335.

50. Haase, M. and Widjajakusuma, J., Damage identification based on ridges and maxima lines of the wavelet transform, *Int. J. Eng. Sci.*, 41, 1423, 2003.

51. Trifunac, M.D., Response envelope spectrum and interpretation of strong earthquake ground motion, *Bull. Seismol. Soc. Am.*, 61, 343, 1971.

52. Spanos, P.D., Donley, M., and Roesset, J. 1987. Evolutionary power spectrum estimation of earthquake accelerograms, September 19, 1985, Mexico. In *Stochastic Approaches in Earthquake Engineering: U.S.–Japan Joint Seminar, Boca Raton, Florida, USA, May 6–7*, Y.K. Lin and R. Minai, Eds., p. 322. Springer, Berlin.

53. Massel, S.R. 1996. *Ocean Surface Waves: Their Physics and Prediction*, World Scientific, River Edge, NJ.

54. Ville, J., Theorie et applications de la notion de signal analytique, *Cables Transm.*, 2, 61, 1948.

55. Loynes, R.M., On the concept of the spectrum for nonstationary processes, *J. R. Stat. Soc. Ser. B Methodol.*, 30, 1, 1968.

56. Gurley, K. and Kareem, A., Applications of wavelet transforms in earthquake, wind, and ocean engineering, *Eng. Struct.*, 21, 149, 1999.

57. Gurley, K., Kijewski, T., and Kareem, A., First- and higher-order correlation detection using wavelet transforms, *J. Eng. Mech.*, 129, 188, 2003.

58. Kareem, A. and Kijewski, T., Time–frequency analysis of wind effects on structures, *J. Wind Eng. Ind. Aerodyn.*, 90, 1435, 2002.

59. Basu, B. and Gupta, V.K., Non-stationary seismic response of multi-DoF systems by wavelet transform, *Earthquake Eng. Struct. Dyn.*, 26, 1243, 1997.

60. Basu, B. and Gupta, V.K., Seismic response of single-DoF systems by wavelet modeling of nonstationary processes, *J. Eng. Mech.*, 124, 1142, 1998.

61. Basu, B. and Gupta, V.K., Stochastic seismic response of single-DoF systems through wavelets, *Eng. Struct.*, 22, 1714, 2000.

62. Basu, B. and Gupta, V.K., Wavelet-based non-stationary response analysis of a friction base-isolated structure, *Earthquake Eng. Struct. Dyn.*, 29, 1659, 2000.

63. Tratskas, P. and Spanos, P.D., Linear multi-DoF system stochastic response by using the harmonic wavelet transform, *J. Appl. Mech.*, 70, 724, 2003.

64. Spanos, P.D. and Failla, G., Evolutionary spectra estimation using wavelets, *J. Eng. Mech.*, 130, 2004.

65. Alkemade, J.A.H. 1993. The finite wavelet transform with an application to seismic processing. In *Wavelets: An Elementary Treatment of Theory and Applications*, T.H. Koornwinder, Ed., p. 183. World Scientific, Hackensack, NJ.

66. Basu, B. and Gupta, V.K., On equivalent linearization using the wavelet transform, *J. Vib. Acoust.*, 121, 429, 1999.

67. Iyama, J. and Kuwamura, H., Application of wavelets to analysis and simulation of earthquake motions, *Earthquake Eng. Struct. Dyn.*, 28, 255, 1999.

68. Massel, S.R., Wavelet analysis for processing of ocean surface wave records, *Ocean Eng.*, 28, 957, 2001.

69. Spanos, P. D., Cacciola, P., and Redhorse, J., Random vibration of SMA systems via Preisach formalism, *Nonlinear Dyn.*, 36, 405, 2004.

70. Priestley, M.B. 1981. *Spectral Analysis and Time Series*, Academic Press, New York.

71. Shinozuka, M. and Deodatis, G., Stochastic process models for earthquake ground motion, *Probab. Eng. Mech.*, 3, 114, 1988.

72. Spanos, P.D. and Mignolet, M.P., Simulation of homogeneous two-dimensional random fields: Part II — MA and ARMA models, *J. Appl. Mech.*, 114, 270, 1992.

73. Fournier, A., Fussell, D., and Carpenter, L., Computer rendering of stochastic models, *Commun. ACM*, 25, 371, 1982.

74. Lewis, J.P., Generalized stochastic subdivision, *ACM Trans. Graph.*, 6, 167, 1987.

75. Fenton, G.A. and Vanmarcke, E.H., Simulation of random fields via local average subdivision, *J. Eng. Mech.*, 116, 1733, 1990.

76. Belkin, G., On the representation of operators in bases of compactly supported wavelets, *J. Numer. Anal.*, 6, 1716, 1993.

77. Robertson, A.N., Park, K.C., and Alvin, K.F., Extraction of impulse response data via wavelet transform for structural system identification, *Proc. Des. Eng. Tech. Conf. DE-84.1 ASME-95*, 1323, 1995.

78. Staszewski, W.J., Identification of damping in multi-DoF systems using time-scale decomposition, *J. Sound Vib.*, 203, 283, 1997.

79. Jones, D.I.G. 1988. Application of damping treatments. In *Shock and Vibration Handbook*, 3rd ed., C.M. Harris, Ed. McGraw-Hill, New York.

80. Staszewski, W.J., Identification of non-linear systems using multi-scale ridges and skeletons of the wavelet transform, *Shock Vib. Dig.*, 214, 639, 1998.

81. Ruzzene, M., Fasana, A., Garibaldi, L., and Piombo, B., Natural frequencies and dampings identification using wavelet transform: application to real data, *Mech. Syst. Signal Process.*, 11, 207, 1997.

82. Lardies, J. and Gouttebroze, S., Identification of modal parameters using the wavelet transform, *Int. J. Mech. Sci.*, 44, 2263, 2002.

83. Spanos, P.D. and Zeldin, B.A., Generalized random decrement method for analysis of vibration data, *J. Vib. Acoust., Trans. ASME*, 120, 806, 1998.

84. Slavic, J., Simonovski, I., and Boltezar, M., Damping identification using a continuous wavelet transform: application to real data, *J. Sound Vib.*, 262, 291, 2003.

85. Lamarque, C.H., Pernot, S., and Cuer, A., Damping identification in multi-DoF systems via a wavelet-logarithmic decrement. Part 1: Theory, *J. Sound Vib.*, 235, 361, 2000.

86. Hans, S., Ibraim, E., Pernot, S., Boutin, C., and Lamarque, C.H., Damping identification in multi-DoF system via a wavelet-logarithmic decrement. Part 2: Study of a civil engineering building, *J. Sound Vib.*, 235, 375, 2000.

87. Ghanem, R. and Romeo, F., A wavelet-based approach for the identification of linear time-varying dynamical systems, *J. Sound Vib.*, 234, 555, 2000.

88. Ghanem, R. and Romeo, F., Wavelet-based approach for model and parameter identification of non-linear systems, *Int. J. Non-Linear Mech.*, 36, 835, 2001.

89. Yu, Y., Xiao, Z., Lin, E.-B., and Naganathan, N., Analytic and experimental studies of a wavelet identification of Preisach model of hysteresis, *J. Magn. Magn. Mater.*, 208, 255, 2000.

90. Mayergoyz, I.D. 2003. *Mathematical Models of Hysteresis and Their Applications*, 1st ed., Academic Press, Boston.

91. Le, T.-P. and Argoul, P., Continuous wavelet transform for modal identification using free decay response, *J. Sound Vib.*, 277, 73, 2004.

92. Yin, H.P., Duhamel, D., and Argoul, P., Natural frequencies and damping estimation using wavelet transform of a frequency response function, *J. Sound Vib.*, 271, 999, 2004.

93. Staszewski, W.J. and Tomlinson, G.R., Application of the wavelet transform to fault detection in a spur gear, *Mech. Syst. Signal Process.*, 8, 289, 1994.

94. Wang, W.J. and McFadden, P.D., Application of orthogonal wavelets to early gear damage detection, *Mech. Syst. Signal Process.*, 9, 497, 1995.

95. Liew, K.M. and Wang, Q., Application of wavelet theory for crack identification in structures, *J. Eng. Mech.*, 124, 152, 1998.

96. Deng, X. and Wang, Q., Crack detection using spatial measurements and wavelet analysis, *Int. J. Fracture*, 91, L23, 1998.

97. Wang, Q. and Deng, X., Damage detection with spatial wavelets, *Int. J. Solids Struct.*, 36, 3443, 1999.

98. Okafor, A.C., Chandrashekhara, K., and Jiang, Y.P., Location of impact in composite plates using waveform-based acoustic emission and Gaussian cross-correlation techniques, *Proc. SPIE — Int. Soc. Opt. Eng.*, 2718, 291, 1996.

99. Hong, J.-C., Kim, Y.Y., Lee, H.C., and Lee, Y.W., Damage detection using the Lipschitz exponent estimated by the wavelet transform: applications to vibration modes of a beam, *Int. J. Solids Struct.*, 39, 1803, 2002.

100. Mallat, S. and Hwang, W.L., Singularity detection and processing with wavelets, *IEEE Trans. Inf. Theory*, 38, 617, 1992.

101. Grossmann, A. 1986. Wavelet transform and edge detection. In *Stochastic Processes in Physics and Engineering*, M. Hazewinkel, Ed. Reidel, Dodrecht.

102. Douka, E., Loutridis, S., and Trochidis, A., Crack identification in beams using wavelet analysis, *Int. J. Solids Struct.*, 40, 3557, 2003.

103. Douka, E., Loutridis, S., and Trochidis, A., Crack identification in plates using wavelet analysis, *J. Sound Vib.*, 270, 279, 2004.

104. Yam, L.H., Yan, Y.J., and Jiang, J.S., Vibration-based damage detection for composite structures using wavelet transform and neural network identification, *Compos. Struct.*, 60, 403, 2003.

105. Moslem, K. and Nafaspour, R., Structural damage detection by genetic algorithms, *AIAA J.*, 40, 1395, 2002.

106. Paget, C.A., Grondel, S., Levin, K., and Delebarre, C., Damage assessment in composites by Lamb waves and wavelet coefficients, *Smart Mater. Struct.*, 12, 393, 2003.

107. Willam, K., Rhee, I., and Beylkin, G., Multiresolution analysis of elastic degradation in heterogeneous materials, *Meccanica*, 36, 131, 2001.

108. Frantziskonis, G., Wavelet-based analysis of multiscale phenomena: application to material porosity and identification of dominant scales, *Probab. Eng. Mech.*, 17, 349, 2002.

109. Frantziskonis, G., Multiscale characterization of materials with distributed pores and inclusions and application to crack formation in an aluminum alloy, *Probab. Eng. Mech.*, 17, 359, 2002.

110. Ciliberto, A., Cavaccini, G., Salvetti, O., and Chimenti, M., Porosity detection in composite aeronautical structures, *Infrared Phys. Technol.*, 43, 139, 2002.

Index

Related Titles

Vibration Simulation Using MATLAB and ANSYS
Michael R. Hatch
ISBN: 1584882050

Vibration: Fundamentals and Practice, Second Edition
Clarence W. de Silva
ISBN: 0849319870